Electrical and Electronic Principles 2

Electrical and Electronic Principles 2

Rhys Lewis B.Sc Tech, C Eng, MIEE

Head of Department of Electronic and Radio Engineering,
Riversdale College of Technology, Liverpool

GRANADA
London Toronto Sydney New York

Granada Publishing Limited – Technical Books Division
Frogmore, St Albans, Herts AL2 2NF
and
36 Golden Square, London W1R 4AH
866 United Nations Plaza, New York, NY 10017, USA
117 York Street, Sydney, NSW 2000, Australia
100 Skyway Avenue, Rexdale, Ontario, Canada M9W 3A6
61 Beach Road, Auckland, New Zealand

Copyright © 1982 Rhys Lewis

ISBN 0 246 11575 0

First published in Great Britain 1982 by Granada Publishing
Printed in Great Britain by William Clowes (Beccles) Limited,
Beccles and London

All rights reserved. No part of this publication may be reproduced,
stored in a retrieval system, or transmitted in any form or by any
means, electronic, mechanical, photocopying, recording or otherwise,
without the prior permission of the publishers.

Granada ®
Granada Publishing ®

Contents

Preface
Self-assessment exercises in this book

1 Circuit theorems
Conductive circuit quantities	1
Ohm's Law	2
Resistors	2
Kirchhoff's Laws	3
Summary	10
Exercise	11
Self-assessment exercise	12
Answers to self-assessment exercise	13

2 Capacitors and capacitance
Electric charge	16
Electric field strength	17
Electrical potential	18
The parallel plate capacitor	20
Capacitors in series and parallel	26
Energy stored in a capacitor	28
Types of capacitor	29
Summary	31
Exercise	32
Self-assessment exercise	34
Answers to self-assessment exercise	35

3 The magnetic field
Basic quantities	38
Magnetic materials	41
Magnetising curves	43
Hysteresis	44
Magnetic circuit quantities and calculations	46
Summary	53
Exercise	54
Self-assessment exercise	55
Answers to self-assessment exercise	57

4 Electromagnetic induction
Force on a current-carrying conductor	61
Electromagnetic induction	63
Self-inductance	65
Mutual inductance	69
The transformer	69
Summary	72
Exercise	73
Self-assessment exercise	74
Answers to self-assessment exercise	76

5 Alternating voltages and currents

	Alternating voltages and currents	78
	Waveform definitions	80
	Graphical addition of alternating voltages and currents	89
	Phasors	91
	Summary	96
	Further worked examples	97
	Self-assessment exercise	99
	Answers to self-assessment exercise	100

6 Single-phase a.c. circuits

	Resistance, reactance and impedance	104
	Series L-R and C-R circuits	109
	Impedance	110
	Series L-C-R circuits	114
	Power in a.c. circuits	118
	Summary	122
	Exercise	124
	Self-assessment exercise	126
	Answers to self-assessment exercise	127

7 Semiconductor diodes and transistors

	Properties of semiconductors	130
	Diodes	134
	Summary	136
	Rectification	136
	Transistors	137
	Using the transistor	142
	Summary	144
	Exercise	145
	Self-assessment exercise	145
	Answers to self-assessment exercise	146

8 Measuring instruments and measurements

	Electromechanical instruments	148
	Errors in measurement using electromechanical instruments	156
	The cathode-ray oscilloscope	161
	Null and bridge methods of measurement	167
	Summary	171
	Exercise	172
	Self-assessment exercise	173
	Answers to self-assessment exercise	175

Preface

This is one of a new series of texts for telecommunications, electronics and electrical engineering students studying for the Technician Education Council courses.

The text covers the unit Electrical and Electronic Principles 2 (U81/747), which was revised in 1981, and increased in length to represent 90 hours teaching. These changes are fully reflected in the book, which is tied to the general and specific objectives of the unit. Each chapter represents a module from the unit, and concludes with a summary and a self-assessment test. Answers are given to all problems and exercises, and each self-assessment test has a self-marking scheme. An additional feature is the inclusion of multi-choice model solutions.

Author's acknowledgements

I would like to thank Tektronix Ltd for permission to include in the text a photograph of one of their oscilloscopes, and I am particularly grateful to Miss Dawn Timmins for the considerable amount of hard work she undertook in the typing and preparation of the manuscript for publication.

Rhys Lewis

Self-assessment exercises in this book

Following each chapter there is a self-assessment exercise which will enable you to test how well you have understood and assimilated the material in the chapter. The layout and marking scheme is similar for each exercise, each one consisting of a number of short-answer or multiple-choice questions followed by long-answer questions. The maximum marks awarded for the short questions are 3 to 5 and for the long ones are 14. Marks are indicated at the side of each question. The maximum for each complete exercise is 100. All questions in each exercise should be attempted.

The time taken for each question and exercise will vary from person to person, of course, but the grade of difficulty of the questions has been carefully chosen so that the maximum time that should be taken is as follows:

    Short questions (3 marks) : 3 minutes
    Short questions (5 marks) : 5 minutes
    Long questions (14 marks) : 30 minutes
    The whole exercise : 3 hours

When the whole exercise has been completed and marked (using the model solutions given), a guide to the level of pass is as follows:

    40–59 : Pass
    60–84 : Pass with merit
    85 and above : Pass with distinction

If the total obtained is below 40 it indicates that an insufficient understanding has been obtained and the chapter should be carefully reworked especially in those areas of difficulty indicated by the marks gained per question.

When marking, a high standard should be adopted in those parts of questions which are subjective (although with the degree of detail given in the marking scheme these have been reduced to a minimum). It is better to apply a high standard to your own marking than to deceive yourself that you have a better understanding than you actually have!

1 Circuit theorems

Topic area: A

General objective
The expected learning outcome is that the student understands circuit theory and applies it to series-parallel circuit problems.

Specific objectives
The expected learning outcome is that the student:
1.1 Applies Ohm's Law to the solution of problems relating to series-parallel combinations of resistors.
1.2 States Kirchhoff's Laws.
1.3 Applies Kirchhoff's Laws to problems involving not more than two unknowns.

Conductive circuit quantities

A conductive circuit is an interconnection of electrical or other components such that electric charge is able to be moved from one point in the circuit to another. Usually, but not always, the charge is carried by electrons. The unit of electric charge is the *coulomb*, symbol C.

Electric current is the rate at which electric charge is moved in a conductive circuit. It is measured in coulombs per second. One coulomb per second (C/s) is called one *ampere*, symbol A.

Voltage is the energy given to or taken from each unit of electric charge as it enters or passes through a conductive circuit. If the energy is being given *to* the charge carriers, from a battery, generator or other source, the voltage is referred to as an *electromotive force* or e.m.f. (It is *not*, however, a force and should not be thought of as such.) If the energy is being taken *from* the charge carriers the voltage is called a *potential difference* or p.d.

Voltage is measured in units of energy per unit charge, that is, joules per coulomb (J/C). One joule per coulomb is called one volt, symbol V.

Power is the rate at which energy is used. It is measured in joules per second (J/s). One joule per second is called one watt, symbol W. Since the unit of voltage is the joule per coulomb and the unit of current is the coulomb per second, if the two units are multiplied together:

$$\text{volts} \times \text{amperes} = \frac{\text{joules}}{\text{coulombs}} \times \frac{\text{coulombs}}{\text{seconds}}$$

$$= \text{joules/second}$$

we obtain the unit of power. Thus at any given instant the product of voltage and current in a circuit or part of a circuit gives the power at that instant.

Ohm's Law

For most conductive materials at constant temperature and in constant environmental conditions the ratio of voltage to current is constant. The ratio is called *electrical resistance* or, simply, resistance since the quantity is a measure of the difficulty in establishing an electric current for a given voltage. The unit of resistance is the volt per ampere (V/A). One volt per ampere is called one ohm, symbol Ω (the Greek letter omega).

If we denote voltage by V, current by I and resistance by R, the usual symbols, then

$$\frac{V}{I} = R, \text{ or } V = IR$$

Figure 1.1

This equation is called Ohm's Law, after the 19th century scientist who investigated the relationship between voltage and current.

Resistors

An electrical component specially designed to have resistance is called a resistor. Resistors may be constructed using various materials and in a variety of different ways. They may be fixed or variable. The symbol for a resistor as used in electrical circuit diagrams is shown in fig. 1.1.

Series and parallel connection

When two or more components are connected so that the whole of the current flowing through one component then flows through the next they are said to be series connected. When two or more components are connected so that the voltage across one component is also applied to the other(s) they are said to be parallel connected. Fig. 1.2 shows two simple circuits consisting of a battery and two resistors.

Note that in these circuits the battery e.m.f. is denoted by E and the p.d. across the components by V. Each of these circuits will be examined in more detail shortly.

When two or more resistors are connected in series the total equivalent resistance is the sum of the individual resistance of each resistor so that for two resistors of resistance R_1 and R_2 (ohms) the total resistance is $R_1 + R_2$. For three resistors of resistance R_1, R_2 and R_3 the total resistance is $R_1 + R_2 + R_3$ and so on.

When two or more resistors are connected in parallel the reciprocal of the total equivalent resistance is the sum of the reciprocals of each individual resistance (the reciprocal of a quantity is one (unity) divided by the quantity). So for two resistors of resistances R_1 and R_2 connected in parallel the total resistance R_{tot} is given by

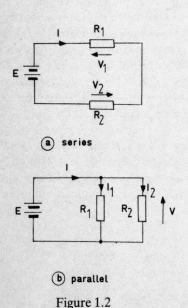

Figure 1.2

$$\frac{1}{R_{tot}} = \frac{1}{R_1} + \frac{1}{R_2}$$

and for three resistors of resistance R_1, R_2 and R_3 connected in parallel

$$\frac{1}{R_{tot}} = \frac{1}{R_1} + \frac{1}{R_2} + \frac{1}{R_3}, \text{ and so on.}$$

The equation for two resistors connected in parallel may be written another way which is easier to remember, for if

$$\frac{1}{R_{tot}} = \frac{1}{R_1} + \frac{1}{R_2}$$

then

$$\frac{1}{R_{tot}} = \frac{R_2}{R_1 R_2} + \frac{R_1}{R_1 R_2}$$

$$= \frac{R_2 + R_1}{R_1 R_2}$$

i.e. $\dfrac{\text{sum of individual resistances}}{\text{product of individual resistances}}$

and $R_{tot} = \dfrac{\text{product of individual resistances}}{\text{sum of individual resistances}}$

but this is true only for two parallel-connected resistances, *not* for more than two.

Kirchhoff's Laws

Let us look again at the simple series circuit shown in fig. 1.2a. It consists of a battery of e.m.f. E volts connected across two resistors of resistance R_1 and R_2 ohms so that a current I amperes flows in the circuit. The voltage across resistor R_1 is shown as V_1 and that across resistor R_2 as V_2.

The battery is providing E joules for each coulomb of charge which is carried round the circuit. This energy is used up by the carriers, electrons in this case, as they move through the conductive leads and the resistors. Energy conversion takes place, in fact, the electrical energy being converted in the main to heat energy. The p.d. across each part of the circuit is a measure of the amount of energy per unit charge used (converted) in that part of the circuit, so that in resistor R_1, for example, V_1 joules are used for each coulomb of charge. In resistor R_2 the number of joules per coulomb used equals V_2.

No other p.d.s are shown since it is assumed that the conductive leads in this circuit do not have resistance and so no energy is required; in practice the amounts will be very small. The whole of the energy supplied per unit charge, i.e. E, is converted as the charge carriers move. The sum of the individual amounts of energy per unit charge used in the various parts of the circuit is $V_1 + V_2$, so that

$$E = V_1 + V_2$$

or alternatively, we can write

$$E - V_1 - V_2 = 0$$

and this is a mathematical expression of Kirchhoff's voltage law which states that:

In any closed loop of a conductive circuit the algebraic sum of the voltages is zero.

By 'the algebraic sum' we mean the sum of the voltages, taking their direction of action into account, i.e. the 'sign' of the voltage. In applying Kirchhoff's voltage law to a circuit or part of a circuit it is essential to indicate clearly at the outset in which direction each voltage is acting.

In circuit diagrams in this book a direct voltage across a component or circuit part will be indicated by an arrow, the arrowhead pointing to the more positive side of the component or circuit part. A closed loop in a circuit is one in which there is a complete path for current flow (although not necessarily the same current, as we shall see later).

In the parallel circuit shown in fig. 1.2b the voltage drop across the parallel-connected resistors is shown as V and is equal to the e.m.f. E (assuming again no voltage loss across leads). However, in this circuit the current supplied by the voltage source does not remain the same throughout the circuit, as in the simple series circuit of fig. 1.2a, but divides as it reaches the junction of R_1 and R_2.

Electric current is a measure of movement of charge and since, in practice, in a conductive circuit such as this one, charge does not accumulate at junctions, the total charge carried away from the junction is equal to the charge carried towards the junction, i.e.

$$I = I_1 + I_2 \text{ in the circuit shown}$$

$$\text{or } I - I_1 - I_2 = 0$$

which is an expression of Kirchhoff's current law, which states:

The algebraic sum of the currents at any junction is zero.

Ohm's Law, the laws governing the total resistance of series and parallel-connected resistors and Kirchhoff's Laws will be used extensively in the remainder of this chapter, which will be devoted to worked examples. These are graded, becoming less simple as the chapter proceeds, and should be studied with care.

Example 1.1 Calculate the total equivalent resistance between points A, B in the circuits shown in fig. 1.3.

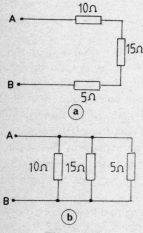

Figure 1.3

(a) In part (a) of the example there are three resistors of value 10 Ω, 15 Ω and 5 Ω connected in series. The total equivalent resistance is therefore $10 + 15 + 5$, i.e. 30 Ω.

(b) In part (b) the resistors are in parallel so that, denoting the equivalent resistance by R_{tot}, we may write

$$\frac{1}{R_{tot}} = \frac{1}{10} + \frac{1}{15} + \frac{1}{5}$$

$$\frac{1}{R_{tot}} = \frac{3 + 2 + 6}{30} = \frac{11}{30}$$

$$\text{and } R_{tot} = \frac{30}{11} = 2.727 \, \Omega$$

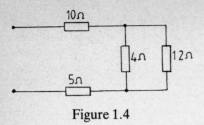

Figure 1.4

Example 1.2 Calculate the equivalent resistance of the circuit shown in fig. 1.4.

The circuit consists of a 10 Ω resistor and a 5 Ω resistor in series with the parallel combination of a 4 Ω and a 12 Ω resistor.

The 4 Ω/12 Ω may be replaced by a single resistance of

$$\frac{4 \times 12}{4 + 12} \; \Omega \text{ (two resistors in parallel)}$$

i.e. 48/16 or 3 Ω

The total resistance is thus $10 + 3 + 5$ ohms, i.e. 18 Ω.

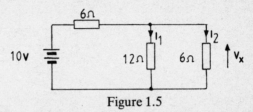

Figure 1.5

Example 1.3 Calculate the current flowing in each branch of the parallel combination in the circuit of fig. 1.5.

To determine the value of I_1 and I_2 we need to know the p.d. across the parallel combination shown as V_x in the circuit.

The total circuit resistance is 6 Ω in series with the parallel combination of 12 Ω and 6 Ω. The 12 Ω/6 Ω combination may be replaced by

$$\frac{12 \times 6}{12 + 6} \text{ i.e. } 4 \; \Omega$$

and the total resistance is thus 6 Ω + 4 Ω, i.e. 10 Ω. The current from the battery is thus 10 V ÷ 10 Ω, i.e. 1 A.

The p.d. across the single 6 Ω resistor is 6 Ω × 1 A, i.e. 6 V (Ohm's Law), and since the e.m.f. is equal to the sum of the p.d.s (Kirchhoff's voltage law),

$$10 = 6 + V_x \text{ and } V_x = 4 \text{ V}$$

The current in the 12 Ω resistor, I_1, is thus 4 V ÷ 12 Ω, i.e. 0.33 A.

The total supply current is 1 A and this current enters the top junction of the parallel combination. Since 0.33 A flows in the 12 Ω resistor then $(1 - 0.33)$ A must flow in the 6 Ω resistor in the parallel combination (Kirchhoff's current law). Thus $I_2 = 1 - 0.33 = 0.67$ A.

Example 1.4 Find the value of currents I_1, I_2 and I_3 when 9 V is applied to the circuit shown in fig. 1.6.

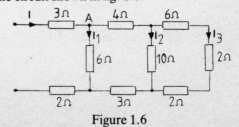

Figure 1.6

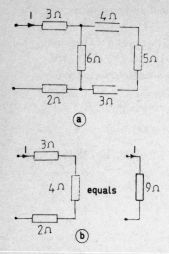

Figure 1.7

To determine each current in turn we require the p.d. across the resistor (or resistors) through which the current flows. To do this we start by determining the total equivalent resistance and the current from the supply and we may then calculate the various p.d.s working from left to right in the circuit. To calculate the equivalent resistance, however, we work from right to left as follows.

The right-hand mesh consists of a 10 Ω resistor in parallel with three series resistors of resistance 6 Ω, 2 Ω and 2 Ω respectively, i.e. 10 Ω in parallel with 10 Ω, which gives an equivalent of 5 Ω.

$$\left(\frac{10 \times 10}{10 + 10}\right) ; \text{see fig. 1.7a}$$

We now have the centre mesh of fig. 1.6 consisting of 6 Ω in parallel with $(4 + 5 + 3)$ Ω, i.e. an equivalent of

$$\frac{6 \times 12}{6 + 12} \text{ or } 4 \, \Omega \, ; \text{see fig. 1.7b}$$

The overall equivalent resistance is then $(3 + 4 + 2)$ Ω, i.e. 9 Ω. The supply current (I in figs 1.6 and 1.7) is then 9 V ÷ 9 Ω, i.e. 1 A.

The p.d. across the 4 Ω resistor in fig. 1.7b and therefore across the 6 Ω resistor and the $(4 + 5 + 3)$ Ω resistors in fig. 1.7a is then $4 \, \Omega \times 1 \, A = 4 \, V$.

The current in the 6 Ω resistor, I_1, is given by

$$I_1 = 4 \, V \div 6 \, \Omega = \frac{4}{6} A = \frac{2}{3} A$$

The current in the $(4 + 5 + 3)$ Ω resistors is $4 \, V \div (4 + 5 + 3)$ Ω, i.e. $\frac{1}{3}$ A, and the p.d. across the 5 Ω resistor in fig. 1.7a and thus the 10 Ω resistor and $(6 + 2 + 2)$ Ω resistors in fig. 1.6 is

$$\frac{1}{3} A \times 5 \, \Omega = \frac{5}{3} V$$

The current in the 10 Ω resistor, I_2, is given by

$$I_2 = \frac{5}{3} \times \frac{1}{10} = \frac{1}{6} A$$

and in the $(6 + 2 + 2)$ Ω resistors (I_3) by

$$I_3 = \frac{5}{3} \times \frac{1}{10} = \frac{1}{6} A$$

(which follows, since $6 + 2 + 2$ is 10 Ω, the value of the other parallel single resistor).

Since $I_2 + I_3$ flows in the 4 Ω resistor of fig. 1.6 the current flowing towards the junction of the 3 Ω, 4 Ω and 6 Ω resistors (point A), I, is equal to $I_1 + I_2 + I_3$, the current flowing away from the junction.

$$I = I_1 + I_2 + I_3 = \frac{2}{3} + \frac{1}{6} + \frac{1}{6} = 1 \, A$$

which we have already determined to be correct. The answers have been checked.

The remaining examples move away from the 'equivalent

Circuit theorems 7

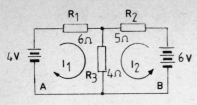

Figure 1.8

resistance' approach and their solutions require the solving of simultaneous equations. The method should be studied carefully.

Example 1.5 Calculate the current in resistor R_3 in the circuit of fig. 1.8.

The current is made up of two meshes shown as A and B. We start by inserting arrows showing the flow of circulating currents *in the direction in which the mesh source (or sources) would drive them*. This is most important if we are to avoid error.

Thus, the 4 V battery drives current I_1 clockwise round the mesh A, flowing through R_1 and R_3. The 6 V battery drives current I_2 anticlockwise round mesh B flowing through R_2 and R_3. Note that R_1 carries I_1 alone, and R_2 carries I_2 alone, but R_3 carries I_1 and I_2 since they flow in the same direction in resistor R_3 (from top to bottom in the figure).

This, of course, is verified by Kirchhoff's current law for the junction of R_1, R_2 and R_3. Currents I_1 and I_2 flow *towards* the junction so the current flowing *away* from the junction (through R_3) is the sum, i.e. $I_1 + I_2$.

We use Kirchhoff's voltage law in turn on each mesh to obtain *two* simultaneous equations to solve for the two unknowns I_1 and I_2. This law states that for each mesh the algebraic sum of the voltages is zero or, alternatively,

$$\text{Sum of e.m.f.s} = \text{sum of p.d.s}$$

For mesh A

$$4 = 6I_1 + 4(I_1 + I_2) \quad \text{equation A}$$

e.m.f. p.d. across R_1 p.d. across R_3

For mesh B

$$6 = 5I_2 + 4(I_1 + I_2) \quad \text{equation B}$$

e.m.f. p.d. across R_2 p.d. across R_3

Equation A gives $4 = 6I_1 + 4I_1 + 4I_2$
$$4 = 10I_1 + 4I_2$$

Equation B gives $6 = 5I_2 + 4I_1 + 4I_2$
$$6 = 4I_1 + 9I_2$$

We must now eliminate one of the unknowns to leave one equation and one unknown. To do this we multiply each equation by an appropriate number such that the coefficients of the unknown we wish to eliminate are the same. (The coefficient is the number preceding the algebraic symbol for the unknown.)

Multiply equation A by 2 to give $8 = 20I_1 + 8I_2$ \hfill (A)

and equation B by 5 to give $30 = 20I_1 + 45I_2$ \hfill (B)

Subtract the new equation A from the new equation B to give

$$30 - 8 = 45I_2 - 8I_2 = 37I_2$$

$$\text{and } I_2 = \frac{22}{37} = 0.595 \text{ A}$$

We may find I_1 either by substitution in one or other of the equations or by using the elimination method. A useful technique is to determine I_1 by elimination of I_2 and use substitution to check the answers.

To find I_1 Multiply equation A by 9 to give $36 = 90I_1 + 36I_2$ (A)

Multiply equation B by 4 to give $24 = 16I_1 + 36I_2$ (B)

Subtract the new equation B from the new equation A to give

$$36 - 24 = 90I_1 - 16I_1 = 74I_1$$

$$I_1 = \frac{12}{74} = 0.162 \text{ A}$$

Check in the original equation A.

$$\text{RHS} = 10I_1 + 4I_2 = (10 \times 0.162) + (4 \times 0.595)$$
$$= 4, \text{ which equals the LHS.}$$

Answer: The current in resistor R_3, $(I_1 + I_2)$, is 0.757 A.

Example 1.6 Find the current in the 3 Ω resistor in fig. 1.9.

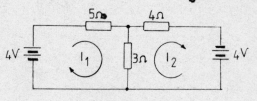

Figure 1.9

Assume two circulating currents I_1, I_2 as shown. Note that each current flows clockwise in the mesh in which it flows so that the required current in the 3 Ω resistor is the difference between I_1 and I_2.

For the left-hand mesh
$$4 = 5I_1 + 3(I_1 - I_2) = 8I_1 - 3I_2$$

For the right-hand mesh
$$4 = 3(I_2 - I_1) + 4I_2 = -3I_1 + 7I_2$$

We now have two equations:

$$4 = 8I_1 - 3I_2$$
$$4 = -3I_1 + 7I_2$$

Multiplying the first throughout by 3 and the second by 8 we obtain

$$12 = 24I_1 - 9I_2$$
$$32 = -24I_1 + 56I_2$$

Adding these equations (to eliminate I_1)

$$44 = 47I_2$$

$$\text{and } I_2 = \frac{44}{47}\text{A} = 0.936 \text{ A}$$

If we now multiply the first of the original equations by 7 and the second by 3 we have

$$28 = 56I_1 - 21I_2$$
$$12 = -9I_1 + 21I_2$$

and adding these (to eliminate I_2),

$$40 = 47I_1$$

$$\text{and } I_1 = \frac{40}{47} = 0.851 \text{ A}$$

Check in the first equation

$$4 = 8I_1 - 3I_2$$
$$\text{RHS} = (8 \times 0.851) - (3 \times 0.936)$$
$$= 4 = \text{LHS}$$

The required current is thus $(0.936 - 0.851)$ A, i.e. 0.085 A, and flows in the direction of I_2, that is, from bottom to top of the 3 Ω resistor in the figure.

Example 1.7 Calculate the current flowing in the 12 Ω resistor in the circuit shown in fig. 1.10.

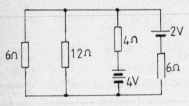

Figure 1.10

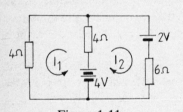

Figure 1.11

To simplify the calculation so that we have only two unknown currents, replace the parallel combination of 6 Ω and 12 Ω by $\frac{6 \times 12}{6 + 12}$, i.e. 4 Ω as in fig. 1.11. Draw in two circulating currents I_1 and I_2 as shown. We may determine how I_1 is divided between 6 Ω and 12 Ω resistors later. Note that to obtain the direction of circulating current flow in the right-hand mesh, in which two batteries are acting, we take the direction in which the battery having the larger e.m.f. would cause current flow.

For the left-hand mesh
$$4 = 4(I_1 + I_2) + 4I_1 = 8I_1 + 4I_2$$

For the right-hand mesh
$$4 - 2 = 4(I_1 + I_2) + 6I_2 = 4I_1 + 10I_2$$

Note that the total *resultant* e.m.f. in this circuit causing circulating current I_2 to flow clockwise is the *difference* between 4 V and 2 V.

The two equations are:
$$4 = 8I_1 + 4I_2$$
$$2 = 4I_1 + 10I_2$$

Doubling the second we have $4 = 8I_1 + 20I_2$, and eliminating I_1 by subtracting the first equation from the new equation: $0 = 16I_2$ and I_2 is zero.

Substituting in the first equation we have $4 = 8I_1$ and $I_1 = 0.5$ A. This checks in the second equation, of which

$$\text{RHS} = (4 \times 0.5) + (10 \times 0) = 2 = \text{LHS}.$$

The 0.5 A flowing into the parallel combination of 6 Ω and 12 Ω divides, the sum of the currents in the branches equalling the current flowing into the combination, according to Kirchhoff's current law. The total equivalent resistance of the combination is 4 Ω, the total current is 0.5 A. The p.d. across the combination is therefore 4×0.5, i.e. 2 V. The current in the 12 Ω resistor is this p.d. divided by the resistance, i.e. $\frac{2}{12}$ or 0.167 A, which is the required answer.

It may appear strange that no current flows in the right-hand 6 Ω resistor. The reason for this is that the batteries are in opposition, 4 V on the one side of the right-hand mesh, 2 V on the other side, and the 4 V is further reduced by the p.d. across the centre 4 Ω resistor (which is 2 V, since 0.5 A is flowing through it). Thus, the net e.m.f. in the right-hand mesh available for circulating a current is $(4 - 2) - 2$, i.e. zero.

Summary Electric current is the movement or flow of electric charge carried usually, but not always, by electrons. The unit of electric charge is the coulomb, symbol C. The unit of current is the ampere, symbol A, which is a flow of 1 coulomb/second.

Electromotive force, abbreviated e.m.f., is the energy given per unit of electric charge by a voltage source; potential difference, abbreviated p.d., is the energy taken from or converted per unit of electric charge when it moves. Both e.m.f. and p.d. are measured in volts, symbol V, where 1 volt is 1 joule/coulomb.

Power is the rate of using energy and is measured in watts, symbol W, where 1 watt is 1 joule/second. The symbols for the quantities current, e.m.f., p.d. and power are, usually, I, E, V and P respectively.

Electrical resistance is the opposition to the flow of electric current. It is measured in ohms, symbol Ω, where 1 ohm is 1 volt/ampere. The symbol for the quantity resistance is, usually, R.

Ohm's Law states that voltage = current × resistance, using symbols $V = IR$, which can be written as $I = V/R$ or $R = V/I$.

A resistor is a component designed to have a particular value of resistance. When resistors are connected together so that the same electric current passes through each in turn they are said to be in series and the total resistance of the series connection is the sum of the individual resistances. For resistors of resistance $R_1, R_2, R_3 \ldots$ etc. the total resistance, R_{tot}, is given by $R_{\text{tot}} = R_1 + R_2 + R_3 + \ldots$ etc.

When resistors are connected together so that the same p.d. exists across each of them they are said to be in parallel. The reciprocal of the total resistance is equal to the sum of the reciprocals of the individual resistances.

For resistors of resistance R_1, R_2, R_3 etc. the total resistance for a parallel connection, R_{tot}, is given by

$$\frac{1}{R_{\text{tot}}} = \frac{1}{R_1} + \frac{1}{R_2} + \frac{1}{R_3} \ldots \text{etc.}$$

For *two* resistances connected in parallel

$$\text{total resistance} = \frac{\text{product of individual resistances}}{\text{sum of individual resistances}}$$

$$\text{i.e. } R_{\text{tot}} = \frac{R_1 R_2}{R_1 + R_2}$$

but this is true only for two resistances, not more than two.

Kirchhoff's Laws concern electric circuits. There are two laws:

The *current law* states that the algebraic sum of the currents at any junction is zero. Simply, this means that the total current entering a junction is equal to the current leaving the junction.

The *voltage law* states that in any closed circuit or part of a circuit the algebraic sum of the voltages in the circuit is zero. Simply, this means that the sum of the e.m.f.s is equal and opposite to the sum of the p.d.s or, if there are no e.m.f.s in part of a circuit but there are p.d.s (as is possible in more complex circuits) some p.d.s will act in one direction, some will act in the other direction and the net p.d. will be zero.

EXERCISE 1

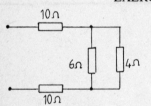

Figure 1.12

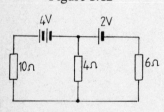

Figure 1.13

For the circuit of fig. 1.12:

1. Find the equivalent total resistance.

2. If the two parallel resistors are replaced by three parallel resistors of equal resistance, calculate the resistance of each parallel resistor for the equivalent total resistance to be the same as before.

3. When a 6 V battery is connected to the circuit, calculate the current in each of the two parallel resistors.

4. If a 6 V battery is connected to the circuit and a 2 V battery is connected in series with the 4 Ω resistor, both batteries having their positive poles uppermost in the figure, calculate the current in the 6 Ω resistor.

For the circuit of fig. 1.13:

5. Calculate the current in the centre limb.

6. If the centre resistor is replaced by a 10 Ω resistor in parallel with a second resistor such that the resistance of the parallel combination remains the same, calculate the current in the second resistor.

7. Calculate the current in the left-hand resistor when the right battery connections are reversed.

8. If the right-hand resistor in the circuit of fig. 1.13 is replaced by the circuit of fig. 1.12, calculate the current flowing in the 4 Ω resistor in the circuit of fig. 1.12.

9. If the centre limb is replaced by a 4 V battery of internal resistance 2 Ω, connected with positive pole downwards in the figure, calculate the current flowing in the new battery.

10. If the right-hand battery is replaced by another such that the currents in each of the outer limbs are of the same value, calculate the e.m.f. of the new battery and state how it is connected.

SELF-ASSESSMENT EXERCISE 1

Possible marks

1. State the units of e.m.f., current and resistance. (3)
2. Define p.d. giving the symbol, unit and unit symbol. (3)
3. State the relationship giving the total resistance R_{tot} of two series-connected resistors, of resistance R_1 and R_2, in terms of R_{tot}, R_1 and R_2. (3)
4. State the relationship giving the total resistance R_{tot} of two parallel-connected resistors R_1 and R_2, in terms of R_{tot}, R_1 and R_2. (3)
5. Two resistors of resistance 4 Ω and 6 Ω are connected in parallel across a 2.4 V supply. The current in the 6 Ω resistor is A. 1 A; B. 0.6 A; C. 0.4 A; D. 0.24 A. (3)
6. Two resistors of resistance 10 Ω and 15 Ω are connected in series across a 30 V supply. Calculate the p.d. across the 10 Ω resistor. (5)
7. Three 10 Ω resistors are connected in parallel across a 10 V supply. Calculate the supply current and the current in each resistor. (5)
8. A 4 Ω resistor is connected in series with the parallel combination of two resistors of resistance 10 Ω and 15 Ω. Calculate the total circuit resistance. (5)
9. Calculate the current I in the following circuit. (14)

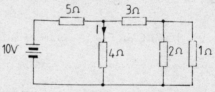

Figure 1.14

10. Calculate the equivalent resistance of the following circuit. (14)

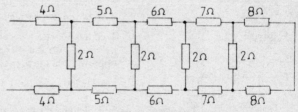

Figure 1.15

11. Calculate the current in the 5 Ω resistor in the following circuit. (14)

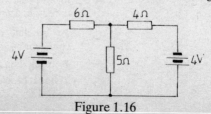

Figure 1.16

12. Calculate the current in the 5 Ω resistor in the circuit of fig. 1.16 if the right-hand battery is reversed. (14)

Circuit theorems 13

13. Calculate the current in the 15 Ω resistor in the following circuit. (14)

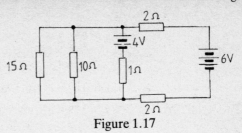

Figure 1.17

Answers

EXERCISE 1

1. 22.4 Ω
2. 7.2 Ω
3. 0.107 A (6 Ω), 0.16 A (4 Ω)
4. 0.2857 A
5. 0.032 A
6. 19.35 mA
7. 0.258 A
8. 20.3 mA
9. 0.75 A
10. 2.4 V positive pole to the left

SELF-ASSESSMENT EXERCISE 1

Marks

1. Volt, ampere and ohm respectively (1) each
2. Potential difference, p.d., is the energy taken from or converted per unit charge (1)
   The quantity symbol is V (1)
   The unit is volt, symbol V (1)
3. $R_{tot} = R_1 + R_2$ (3)
4. $\dfrac{1}{R_{tot}} = \dfrac{1}{R_1} + \dfrac{1}{R_2}$

   or $R_{tot} = \dfrac{R_1 R_2}{R_1 + R_2}$ (3)

5. C is the correct answer, obtained by dividing 2.4 V by 6 Ω. The same (supply) voltage is across each resistor since they are in parallel. A is the total circuit current, B is the current in the 4 Ω resistor (2.4 V/4 Ω). D is obtained by dividing 2.4 V by 10 Ω which would be the resistance if the resistors were in series. This answer would then be the circuit current. (3)

6. Circuit resistance = 10 + 15 = 25 Ω (2)
   Circuit current = $\dfrac{30}{25}$ = 1.2 A (2)
   P.D. across 10 Ω resistor = 10 × 1.2 = 12 V (1)

7. The 10 V supply is across each resistor since they are in parallel.
   Current in each resistor = $\dfrac{10\text{ V}}{10\text{ Ω}}$ (3)
   Total circuit current = 1 + 1 + 1 = 3 A (2)

8. Resistance of parallel combination

$$\frac{10 \times 15}{10 + 15}, \text{ i.e. } 6\,\Omega \qquad (2)$$

which is in series with $4\,\Omega$ to give a total circuit resistance of $4\,\Omega + 6\,\Omega$, i.e. $10\,\Omega$. (3)

9. Equivalent resistance of $1\,\Omega$ and $2\,\Omega$ in parallel is $\frac{1 \times 2}{1+2}$, i.e. $0.67\,\Omega$. (2)

This is in series with $3\,\Omega$ to give a resistance $3 + 0.67\,\Omega$, i.e. $3.67\,\Omega$. (2)

This is in parallel with $4\,\Omega$ to give $\frac{4 \times 3.67}{4 + 3.67}$, i.e. $1.91\,\Omega$. (2)

This is in series with $5\,\Omega$ to give $5 + 1.91$, i.e. $6.91\,\Omega$. (2)

Equivalent circuit resistance is then $6.91\,\Omega$ and circuit current is $10/6.91$, i.e. $1.45\,A$. This current flows in the $5\,\Omega$ resistor and in the parallel combination of $4\,\Omega$ and $3.67\,\Omega$ (the equivalent resistance of the rest of the circuit), i.e. the $1.91\,\Omega$ resistance. (2)

P.d. across the parallel combination $1.91\,\Omega$ is 1.45×1.91 V i.e. 2.77 V, and the current in the $4\,\Omega$ resistor is this p.d. divided by 4, i.e. $\frac{2.77}{4}$ which equals $0.69\,A$. (2) (2)

10. Starting at the right-hand side of the circuit:

$(8\,\Omega + 8\,\Omega)$ in parallel with $2\,\Omega$ gives an equivalent resistance of $\frac{16 \times 2}{16 + 2}$, i.e. $1.78\,\Omega$. (3)

This is in series with $(7 + 7)\,\Omega$ to give $15.78\,\Omega$ which is in parallel with $2\,\Omega$ to give $\frac{2 \times 15.78}{2 + 15.78}$, i.e. $1.77\,\Omega$. (3)

This is in series with $(6 + 6)\,\Omega$ to give $13.77\,\Omega$ which is in parallel with $2\,\Omega$ to give $\frac{2 \times 13.77}{2 + 13.77}$, i.e. $1.75\,\Omega$. (3)

This is in series with $(5 + 5)\,\Omega$ to give $11.75\,\Omega$, which is in parallel with $2\,\Omega$ to give $\frac{2 \times 11.75}{2 + 11.75}$, i.e. $1.71\,\Omega$. (3)

This is in series with $(4 + 4)\,\Omega$ to give $9.71\,\Omega$, which is the total equivalent resistance of the circuit. (2)

11. Let the circulating current in the left-hand mesh be I_1. It flows clockwise. Let the circulating current in the right-hand mesh be I_2. It too flows clockwise.

Left-hand mesh: $4 = 6I_1 + 5I_1 - 5I_2 = 11I_1 - 5I_2$ (3)

Right-hand mesh: $4 = 5I_2 - 5I_1 + 4I_2 = -5I_1 + 9I_2$ (3)

First equation $\times 5$ gives $20 = 55I_1 - 25I_2$

Second equation $\times 11$ gives $44 = -55I_1 + 99I_2$

Add: $64 = 74I_2$ and $I_2 = 0.865\,A$. (2)

Substitute in $4 = 11I_1 - 5I_2$ to give $4 = 11I_1 - (5 \times 0.865)$, $11I_1 = 8.32$ and $I_1 = 0.757\,A$. (2)

Check these values in $4 = -5I_1 + 9I_2$
RHS $= (-5 \times 0.757) + (9 \times 0.865) = 4 =$ LHS. (2)

Current in $5\,\Omega$ resistor is $I_2 - I_1$ flowing from bottom to top, i.e. $0.865 - 0.757$ which equals $0.108\,A$. (2)

12. When the right-hand battery is reversed, using the notation as in the previous question, but remembering that the right-hand mesh current is now *anticlockwise* (i.e. the current in the $5\,\Omega$ resistor is the *sum* of I_1 and I_2) the equations are

Left-hand mesh: $4 = 6I_1 + 5I_1 + 5I_2 = 11I_1 + 5I_2$ (3)

Right-hand mesh: $4 = 4_2 + 5I_2 + 5I_1 = 5I_1 + 9I_2$ (3)

First equation × 5 gives $20 = 55I_1 + 25I_2$

Second equation × 11 gives $44 = 55I_1 + 99I_2$

Second equation minus first equation gives $24 = 74I_2$ and $I_2 = 0.32$ A. (2)

Substitute in $4 = 11I_1 + 5I_2 = 11I_1 + (5 \times 0.32)$: $I_1 = 0.22$ A. (2)

Check in $4 = 5I_1 + 9I_2$
RHS $= (5 \times 0.22) + (9 \times 0.32) = 4 =$ LHS. (2)

Current in 5 Ω resistor is the *sum* of I_1 and I_2 i.e. 0.22 A + 0.32 A which equals 0.54 A. (2)

13. Replace the 15 Ω/10 Ω parallel combination by its equivalent of $\frac{15 \times 10}{15 + 10}$, i.e. 6 Ω (2)

Let the circulating current in the left-hand mesh be I_1 (anticlockwise) and in the right-hand mesh be I_2 (also anticlockwise). The equations are:

Left-hand mesh: $4 = 6I_1 + I_1 - I_2$

Right-hand mesh: $6 - 4 = 2I_2 + 2I_2 + I_2 - I_1$

which gives $4 = 7I_1 - I_2$ (3)
and $2 = -I_1 + 5I_2$ (3)

First equation × 5: $20 = 35I_1 - 5I_2$
Add to second equation: $22 = 34I_1$ and $I_1 = 0.65$ A. (2)

The current flows in the 6 Ω which replaces the parallel combination of 10 Ω and 15 Ω. The current in the 15 Ω resistor is required.
P.D. across 6 Ω = 6 × current = 6 × 0.65 = 3.9 V (2)

Current in 15 Ω resistor = 3.9/15 A = 0.26 A (2)

Note that the value of I_2 is not required. The value of I_1 may also be found by first determining I_2 and using this to find I_1. The process is longer and unnecessary.

2 Capacitors and capacitance

Topic area: B

General objective The expected learning outcome is that the student understands the concept of capacitance and solves problems involving capacitors.

Specific objectives The expected learning outcome is that the student:
2.1 States that charged bodies attract or repel each other.
2.2 Expresses field strength as force per unit charge.
2.3 Defines potential and potential difference.
2.4 Expresses field strength as potential gradient.

A *capacitor* is an electrical component which is capable of storing electric charge after a voltage has been applied across its terminals. The amount of electric charge stored for each volt applied is called the *capacitance* of the capacitor. Capacitance depends upon a number of physical characteristics of the capacitor including size, type of materials used and so on. The process of charging and discharging a capacitor affects voltage levels in a circuit and in addition to charge storage the component may also be used for a number of purposes when alteration of voltage waveforms is required.

Electric charge All matter is made up of atoms. Each atom in turn contains electrons, protons and neutrons, which may be thought of as tiny particles. It is found that a force exists between these particles and that electrons repel other electrons but attract protons and protons repel other protons but attract electrons. Neutrons do not appear to set up a force. We say that both electrons and protons have electric charge and that there are two kinds, positive and negative. The electron has negative charge and the proton has positive charge.

It is possible to treat certain materials so that they too show the properties of electric charge and set up a force. For example, when ebony is rubbed with fur it acquires a charge similar to that on an electron. When glass is rubbed with silk it acquires a charge similar to that on a proton. Uncharged materials have equal numbers of electrons and protons in their atomic structure and, since the charge on an electron is equal and opposite to that on a proton, there is no resultant charge. Friction transfers electrons from one material to another and the material gaining electrons becomes negatively charged; the material losing electrons becomes positively charged. Touching an uncharged material with a charged one can also transfer electrons and the uncharged material may become charged.

Charged bodies attract or repel each other and attract uncharged bodies, so that a negatively charged body will attract a positively charged body but repel another negatively charged body in a similar way to the electrons and protons mentioned earlier. Charged bodies having either kind of charge attract uncharged bodies and this is easily demonstrated. When a comb is used it becomes charged due to friction between it and the hair and the comb will then attract and pick up tiny pieces of paper. The usual demonstration of the effects of charge uses a very light material called pith, the lining of tree bark, and the various forces are easily shown when pieces of pith are charged by touching with other charged bodies.

The basic rule concerning forces set up by charged bodies is that

> like charges repel
> unlike charges attract

and charged bodies attract uncharged bodies.

Electric field strength

In the region surrounding a charged body in which forces can be felt we say that the body sets up an *electric field* and we measure the strength of this field in terms of the force experienced by another body carrying one coulomb of charge placed in the field. Thus if the body carrying unit charge experiences a force of N newtons we say the field strength is N newtons per coulomb (N/C).

Electric field strength of an electric field is a measure of the force acting on a body carrying one coulomb of charge when it is placed in the field. It is measured in newtons per coulomb and has the symbol E.

Electric flux and electric flux density

When drawing diagrams to illustrate electric fields lines are drawn showing the direction of action of the force acting on a positively charged body placed in the field. These lines are called lines of *electric flux*, or flux for short. One flux line is drawn for each coulomb of charge on the body which has set up the field so that if its charge is Q coulombs, the electric flux associated with it has Q lines and we sometimes refer to 'coulombs of flux' although, strictly speaking, the coulomb is the unit of charge and flux is an idea or concept rather than a measurable physical quantity. If the figure representing electric flux (i.e. the charge producing it) is divided by the area in which the field is experienced we obtain *electric flux density*, symbol D, measured in coulombs/square metre. The area is the cross-section through which lines of flux are drawn at right angles. See fig. 2.1.

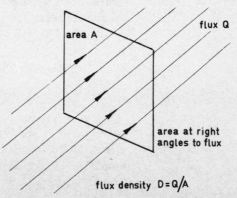

Figure 2.1

Electrical potential

When a charged body is placed in an electric field it experiences a force and if it is able to move it will do so. Thus, it has energy by virtue of its position, in the same way that a body held above ground in the gravitational field of the earth has. Such energy is called potential energy, for if the body is released this energy is converted to other kinds of energy as the body accelerates.

In the electric field we are interested in the potential energy of a charged body at a point in the field and in this case we talk of 'electrical potential energy', 'electrical potential' or *potential* for short. Electrical potential at a point in an electric field is measured as the potential energy of a body having unit charge when it is placed at that point in the field. The unit of measurement is the unit of energy (joule) per unit of charge (coulomb) and, as we have seen before, there is a special name given to a joule per coulomb. It is the *volt*, symbol V.

Electrical potential at a point in an electric field is the potential energy of a body carrying unit charge when placed at that point in the field. It is measured in joules/coulombs or volts. The symbol for electrical potential is V.

Potential difference

Potential energy is energy because of position and thus the electrical potential of a charged body in an electric field depends upon its position in the field. The potential at different points in the field may be different and we then talk of a difference in potential or potential difference (p.d.) between points. The unit of potential difference is, of course, the same as the unit of potential, the volt.

Potential difference between two points in an electric field is the difference between the electrical potential at the points. It is measured in volts and has the symbol V.

The potential difference we meet here in the study of non-moving or static charged bodies is, of course, exactly the same quantity we met in our earlier studies in the conductive circuit. In the conductive circuits we have discussed, the electric field is set up not by bodies that have been charged by friction or by touch but by chemical means in the case of a battery (or electromagnetic means in the case of a generator).

In this case we talked of a 'voltage source' but as can now be seen, the source actually establishes an electric field and in the conductive circuit case charged bodies actually move in the field. Equally a voltage source can be used to set up a field in which charged bodies do not move and this is the basic principle of operation of a capacitor.

Voltage gradient

Whether or not current flows in a material in which an electric field is set up depends upon the material itself and on the strength of the electric field. Any material, even one normally considered to be an insulator, will break down and conduct if a sufficiently strong electric field is established in it. If the electric field is one which is set up using a voltage source the strength of the field obviously depends upon the applied voltage.

It also, however, depends upon the thickness of material across which the voltage is applied. As an example a p.d. of, say, 0.5 V does not sound excessive but if it is applied across a piece of material 0.01 mm thick, it is equivalent to applying 50 kV across a piece 1 m thick. These figures suggest a strong field. When the voltage across a material is divided by the material thickness the quantity obtained is called *voltage gradient*, measured in volts/metre. If we look more closely at the units of voltage gradient we see that

$$\frac{\text{volt}}{\text{metre}} = \frac{\text{joule}}{\text{coulomb}} \times \frac{1}{\text{metre}}$$

$$= \frac{\text{newton-metre}}{\text{coulomb-metre}}$$

$$= \frac{\text{newton}}{\text{coulomb}}$$

which is the unit of electric field strength. Voltage gradient and electric field strength are thus the same quantity.

The value of the voltage gradient at which breakdown occurs is called the *dielectric strength* of the dielectric. If the voltage gradient across a capacitor dielectric exceeds the dielectric strength, conduction takes place in the dielectric and the capacitor is made useless. Some typical values in MV/m are as follows:

    Air   2
    Glass 15
    Mica   100
    Paper 5

Dielectric strength $= \dfrac{\text{breakdown voltage}}{\text{dielectric thickness}}$

and thus breakdown voltage = dielectric strength × dielectric thickness

so that an air dielectric capacitor having a plate separation of 1 mm, for example, would break down if the voltage across it exceeded

$$2 \times 10^6 \times 1 \times 10^{-3}, \text{i.e. } 2000 \text{ V}$$

Capacitors are normally marked to show the safe working voltage. The value given may be in 'volts d.c.' or 'volts a.c.', the latter meaning the r.m.s. value of the alternating voltage. A capacitor marked in volts d.c. should not be subjected to an alternating voltage having a *peak* value greater than the direct voltage indicated, otherwise breakdown will occur.

Specific objectives

The expected learning outcome is that the student:
2.5 States that charge Q on an object is proportional to its potential.
2.6 Defines capacitance as Q/V.
2.7 States the unit for capacitance as the farad.

2.8 States that capacitance of a parallel plate capacitor is proportional to A, inversely proportional to d and depends upon the medium between the plates.
2.9 States that there is a uniform field between parallel plates and that its strength is V/d.
2.10 Defines dielectric constant.
2.13 Relates dielectric strength to capacitor working voltage.

The parallel plate capacitor

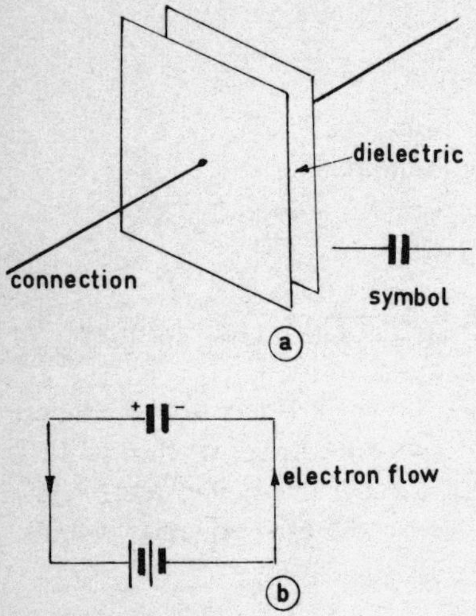

Figure 2.2

A simple parallel plate capacitor consists of two conductive plates separated by a material of uniform thickness as shown in fig. 2.2a. The separating material is called the capacitor *dielectric*. Under normal conditions the dielectric should not conduct so that it is usually made of an insulating material, some common examples being air, paper, mica, ceramic and polystyrene.

When this simple capacitor is connected to a voltage source as shown in fig. 2.2b, electrons move from the negative pole of the battery to the right-hand plate of the capacitor which then becomes negatively charged. A uniform electric field is set up across the dielectric repelling electrons from the left-hand plate. These electrons leave the plate and are attracted to the positive pole of the battery.

The process is temporary or *transient* since as the right-hand plate becomes progressively more negatively charged a force of repulsion tends to slow down and eventually stop the movement of further electrons on to the plate. After the transient period there is no further movement of electrons and the capacitor right-hand plate is fully charged negatively, the left-hand plate fully charged positively (since it has lost electrons).

If the voltage source is removed and the capacitor is situated in a dry insulating medium it will hold its charge indefinitely. Connection of a wire between the plates discharges the capacitor as the extra electrons on the negative plate move through the wire to the positive plate until both plates have zero charge once again.

The total amount of charge acquired by the capacitor is directly proportional to the voltage applied across the plates and we can write

$$Q \propto V$$

where Q represents charge (coulombs) and V the applied voltage (volts).

A statement of proportionality such as this may be written as an equation by inserting a constant:

$$Q = CV$$

The constant C is called the *capacitance* of the capacitor. It is obtained by dividing the acquired charge by the applied voltage, its units being coulombs per volt.

Capacitance of a capacitor is the charge stored per unit voltage. It is measured in coulombs per volt. One coulomb/volt is given the special name the farad, symbol F. The symbol for capacitance is C.

As with all units, multiples and sub-multiples may be used, the microfarad (μF), picofarad (pF) and nanofarad (nF) being common examples.

$$1\,\mu\text{F} = 1 \times 10^{-6}\,\text{F}$$
$$1\,\text{pF} = 1 \times 10^{-12}\,\text{F}$$
$$1\,\text{nF} = 1 \times 10^{-9}\,\text{F}$$

Example 2.1 Calculate the charge stored in a 0.1 μF capacitor when 100 V has been applied for a time sufficient to allow the capacitor to be fully charged. If the dielectric thickness is 0.1 mm determine the electric field strength.

Charge = capacitance × voltage
$$= 0.1 \times 10^{-6} \times 100$$
$$= 10\,\mu\text{C}$$

Electric field strength = voltage/dielectric thickness

$$= \frac{100}{0.1 \times 10^{-3}}$$

$$= 1\,\text{MV/m (note the units)}$$

Effect of plate area and separation

The larger the area of the capacitor plates the greater is the amount of charge stored per unit voltage applied, that is, the capacitance. It is found that, in fact, capacitance is *directly proportional* to plate area, all other factors remaining the same, so that doubling the area doubles the capacitance, trebling the area trebles the capacitance and so on. Denoting plate area by A (square metres) and capacitance by C (farads)

$$C \propto A$$

The plate separation, which is, of course, the thickness of the dielectric separating the plates, also affects capacitance. For a particular voltage applied, increasing the separation *reduces* the electric field strength (since this equals voltage/separation) and thus the charge stored per volt, i.e. the capacitance. All other things being equal the capacitance is, in fact, indirectly (or inversely) proportional to plate separation, so that doubling the separation halves the capacitance, halving the separation doubles the capacitance and so on. Denoting plate separation by d (metres) and capacitance by C (farads)

$$C \propto \frac{1}{d}$$

and we may put both together and write

$$C \propto \frac{A}{d}$$

This may be written as an equation by including an appropriate constant, as we shall see later.

Example 2.2 Two capacitors have the same dielectric material. In one capacitor the plate area is twice that of the other and the plate separation is half that of the other. The capacitor having the smaller plate area stores 1 mC of charge when fully charged after 50 V has been applied across its plates. Calculate the capacitance of the capacitor having the larger plate area.

The capacitor with the smaller plate area stores 1×10^{-3} C after 50 V has been applied. If C farads is its capacitance,

$$1 \times 10^{-3} = C \times 50$$
(charge = capacitance × voltage)

$$C = \frac{10^{-3}}{50} \text{ F}$$

$$= \frac{1000}{50} \times 10^{-6} = 20 \text{ }\mu\text{F}$$

The other capacitor has a plate area twice the size of this one; its capacitance would therefore *increase* two times. Its plate separation is half that of the other and this would cause a further twofold increase ($1/\frac{1}{2}$ since capacitance is indirectly proportional to plate separation). The total increase is thus 2×2, i.e. 4, and its capacitance is therefore 4×20 μF.

The capacitance of the capacitor having the larger plate area is 80 μF.

Example 2.3 Calculate the voltage across a 10 μF capacitor when the charge on the capacitor is 100 μC.

$$\text{Voltage} = \text{charge/capacitance}$$
$$= 100 \times 10^{-6}/10 \times 10^{-6} = 10 \text{ V}$$

Example 2.4 If the plate area and the dielectric thickness of a capacitor are doubled, the capacitance:

    A. remains the same
    B. is halved
    C. is doubled
    D. is divided by four

Since capacitance is directly proportional to plate area and inversely proportional to dielectric thickness (plate separation) it would be doubled (×2) by doubling the area but halved ($\frac{1}{2}$) by doubling the separation. The overall effect is a change of 2 (area) $\times \frac{1}{2}$ (separation) which is unity, i.e. there is no change and A is the correct answer.

Answer B ignores the effect of the area, answer C ignores the effect of plate separation and answer D is obtained by supposing the capacitance is directly proportional to separation *and* indirectly proportional to area, which is, of course, incorrect.

Example 2.5 The charge on a 0.5 μF capacitor when 10 V is applied is equal to:

A. $0.05\,\mu C$
B. $20\,MC$
C. $5\,C$
D. $5\,\mu C$

Since charge = capacitance × voltage, the answer is $0.5 \times 10^{-6} \times 10$, i.e. 5×10^{-6} C or $5\,\mu C$. Answer D is correct. Answer C neglects the multiplier 10^{-6} and is incorrect. Answer A is obtained by dividing capacitance by voltage, answer B by dividing voltage by capacitance. Correctly, capacitance and voltage are *multiplied* to give charge.

The effect of the dielectric: dielectric constant

The material between the plates of a capacitor, the dielectric, affects the electric field between the plates and thus the charge stored per volt applied, the capacitance. Some materials readily support an electric field, others not as readily, in a similar manner that some materials support electric current in a conductive circuit better than others.

We have seen that capacitance is directly proportional to plate area and indirectly proportional to plate separation. Using the symbols C, A and d for capacitance (F), area (m²) and separation (m) respectively, we can write

$$C \propto \frac{A}{d}$$

To write an equation connecting these quantities, a constant is inserted into the statement of proportionality. Using the Greek letter ε (epsilon) for the constant, the equation is

$$C = \varepsilon \frac{A}{d}$$

The effect of the dielectric is indicated by the value of the constant ε which has the same value for a particular dielectric regardless of the values of A and d. If the dielectric is changed the constant ε has a new value.

Called the *absolute permittivity* of the dielectric, ε is a measure of how well an electric field is supported by the dielectric per unit voltage applied across the capacitor plates. If the plates were one metre square and the plate separation were one metre (an impractical case but the assumption is useful as we shall see),

$$A = 1\,m^2 \text{ and } d = 1\,m$$

$$\text{then } C = \varepsilon \frac{1}{1} = \varepsilon$$

and we see that the absolute permittivity of a dielectric is the capacitance between plates of unit area separated by unit distance or, looking at it another way, the capacitance between the opposite faces of a cube of side one metre. See fig. 2.3.

This is in fact one way in which absolute permittivity is defined.

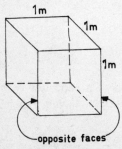

Figure 2.3

Practical measurement is not achieved by constructing such a capacitor, however!

The units of absolute permittivity may be obtained by examination of the general formula for capacitance:

$$C = \varepsilon \frac{A}{d} \text{ using the symbols as before}$$

and so $\varepsilon = \dfrac{Cd}{A}$, the units of which are

$$\frac{\text{farads} \times \text{metres}}{\text{metres}^2} \text{ or farads/ metre (F/m)}$$

The absolute permittivity, ε, of a material is the capacitance between opposite faces of a cube of the material having sides each of length one metre. Its units are farads/metre.

The absolute permittivity of a vacuum is known as the *permittivity of free space* and has the value 8.85×10^{-12} F/m. It is given the symbol ε_0. The absolute permittivity of any other dielectric is usually expressed in terms of the permittivity of free space by using a second constant called the *relative permittivity*, symbol ε_r.

Absolute permittivity = relative permittivity × permittivity of free space.

$$\varepsilon = \varepsilon_r \varepsilon_0$$

so that the general equation, using the symbols as before, now becomes

$$C = \frac{\varepsilon_r \varepsilon_0 A}{d}$$

Relative permittivity is also called the *dielectric constant*.

The relative permittivity or dielectric constant of a material, ε_r, is the ratio between the absolute permittivity of the material and the permittivity of free space. Since it is a ratio it has no units.

Some examples of materials used as capacitor dielectrics and typical values of their relative permittivities are:

Dry air	1
Polyester	3
Paper	4.5
Mica	6
Tantalum oxide	25
Ceramic	10 000

Example 2.6 Calculate the capacitance of a capacitor of plate area 6.25×10^{-4} m², plate separation 5 mm and having a dielectric of relative permittivity 4.

$$\text{Capacitance} = \frac{\text{absolute permittivity} \times \text{plate area}}{\text{plate separation}}$$

Absolute permittivity = relative permittivity × permittivity of free space

and in this case $\varepsilon = 4 \times 8.85 \times 10^{-12}$

so that capacitance $= \dfrac{4 \times 8.85 \times 10^{-12} \times 6.25 \times 10^{-4}}{5 \times 10^{-3}}$

$= 44.25 \times 10^{-13}$ F $= 4.425$ pF

(Note that the plate separation must be in metres)

Example 2.7 Calculate the ratio plate area/plate separation of a 0.5 μF capacitor using dry air as its dielectric (ε_r for dry air may be taken as unity).

From the general equation for capacitance,
$$\dfrac{\text{plate area}}{\text{plate separation}} = \dfrac{\text{capacitance}}{\text{absolute permittivity}}$$
and for dry air, absolute permittivity $= 1 \times 8.85 \times 10^{-12}$ so that the ratio in this case
$$\dfrac{\text{plate area}}{\text{plate separation}} = \dfrac{0.5 \times 10^{-6}}{8.85 \times 10^{-12}} = 56.5 \times 10^3$$

This figure indicates that for a separation of, say, 5 mm the plate area must be $56.5 \times 10^3 \times 5 \times 10^{-3}$, i.e. 282.5 m^2. For a separation of 0.5 mm the area must be 28.25 m^2, for a separation of 0.05 mm the area must be 2.825 m^2 (2825 cm^2) and we see that the area is still fairly large even when a very small separation is used.

To overcome the need for large area plates, multiplate capacitors are constructed such that the total plate area is the sum of the area of several plates, each of much smaller area. The plates are connected together internally so that the number of capacitor connections may remain at two.

Example 2.8 A capacitor of plate area 0.02 m^2 and plate separation 15 μm has a charge of 250 nC when 150 V is applied.

Calculate the:
(a) electric field strength between the plates,
(b) capacitance of the capacitor,
(c) relative permittivity of the dielectric.

(a) Electric field strength = voltage/plate separation
$= 150/(15 \times 10^{-6}) = 10$ MV/m

(b) Capacitance = charge × voltage
$= 250 \times 10^{-9} \times 150 = 37\,500 \times 10^{-9}$ F $= 37.5$ μF

(c) The relative permittivity may be obtained from the equation
$$C = \dfrac{\varepsilon_r \varepsilon_0 A}{d}$$
so that $\varepsilon_r = \dfrac{C d}{\varepsilon_0 A}$

and in this case $\varepsilon_r = \dfrac{37.5 \times 10^{-6} \times 15 \times 10^{-6}}{8.85 \times 10^{-12} \times 0.02} = 3178$

Specific objectives

The expected learning outcome is that the student:

2.11 Calculates the equivalent capacitance of capacitors connected in series, and connected in parallel.

2.12 Solves simple problems involving series-parallel capacitors.

Capacitors in series and parallel

Capacitors may be connected in series or in parallel as shown in fig. 2.4.

When a voltage is applied to two or more capacitors connected in series the charge on each capacitor is the same since the same transient current flows in the leads connecting the capacitors for the same amount of time (electric current is charge per unit time). The voltage on each capacitor adjusts so that the relationship charge = capacitance × voltage is true for each capacitor, the higher the capacitance the lower the capacitor voltage and vice versa.

For the circuit shown in fig. 2.4a:

Total voltage = sum of individual capacitor voltages

$$V = V_1 + V_2 + V_3$$

For each capacitor, capacitor voltage = charge/capacitance, and for the circuit as a whole, total voltage = charge/equivalent capacitance. Denoting the equivalent capacitance by C_E we have:

$$\frac{Q}{C_E} = \frac{Q}{C_1} + \frac{Q}{C_2} + \frac{Q}{C_3}$$

and dividing throughout by Q,

$$\frac{1}{C_E} = \frac{1}{C_1} + \frac{1}{C_2} + \frac{1}{C_3}$$

We see that the reciprocal of the equivalent capacitance is equal to the sum of the reciprocals of the individual capacitances. For two capacitances:

$$\frac{1}{C_E} = \frac{1}{C_1} + \frac{1}{C_2} = \frac{C_2 + C_1}{C_1 C_2}$$

and $C_E = C_1 C_2/(C_1 + C_2)$, i.e. product/sum (similar to the result for resistors in *parallel*).

When capacitors are connected in parallel as shown in fig. 2.4b, the voltage across each capacitor is the same so that the charge on each capacitor adjusts according to the relationship charge = capacitance × voltage.

Denoting the charge on each capacitor by Q_1, Q_2, Q_3, etc. we can say that the total charge

$$Q = Q_1 + Q_2 + Q_3$$

and since total charge = equivalent capacitance × voltage

$$C_E V = C_1 V + C_2 V + C_3 V$$
$$\text{and } C_E = C_1 + C_2 + C_3$$

and we see that the equivalent capacitance of capacitors connected in parallel is the sum of the individual capacitances. (Similar to the result for resistors connected in *series*.)

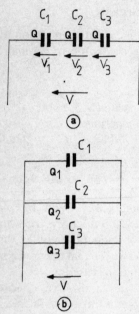

Figure 2.4

Example 2.9 Three 0.5 μF capacitors are connected in series and the combination placed in parallel with a 0.1 μF capacitor across a 100 V supply. Calculate the total charged stored by the circuit.

Denoting the equivalent capacitance of the series circuit by C_E (μF) we have

$$\frac{1}{C_E} = \frac{1}{0.5} + \frac{1}{0.5} + \frac{1}{0.5} = \frac{3}{0.5} \text{ and } C_E = \frac{0.5}{3}, \text{ i.e. } 0.167 \text{ μF}$$

This equivalent capacitance is in parallel with 0.1 μF so that the equivalent circuit capacitance is (0.167 + 0.1) μF, i.e. 0.267 μF.

Using charge = capacitance × voltage, we have

Charge = $0.267 \times 10^{-6} \times 100 = 26.7$ μC

Example 2.10 The equivalent capacitance of a certain circuit is 5 μF. The circuit is made up of two series-connected capacitors of equal value connected across a 2.5 μF capacitor. Determine the value of each of the series-connected capacitors and the p.d. required across the circuit so that 50 μC of charge would be stored in the series branch of the circuit.

Since we add the capacitances of parallel-connected capacitors to obtain the equivalent capacitance, we have:

5 = 2.5 + equivalent capacitance of series branch (μF), so that the equivalent capacitance of series branch = 2.5 μF.

Denoting the capacitance of each series capacitor by C_S (μF) we can write:

$$\frac{1}{2.5} = \frac{1}{C_S} + \frac{1}{C_S} = \frac{2}{C_S} \text{ and } C_S = 2 \times 2.5 = 5 \text{ μF}$$

The total capacitance of the series branch = 2.5 μF
Charge stored in the series branch = 50 μC, so that 50×10^{-6} = 2.5×10^{-6} × voltage required

$$\text{and voltage required} = \frac{50 \times 10^{-6}}{2.5 \times 10^{-6}} = 20 \text{ V}$$

Example 2.11 Six components are connected in series. Three are 33 Ω resistors, three are 0.1 μF capacitors. The equivalent resistance and capacitance of this circuit are, respectively:

A. 99 Ω, 0.3 μF; B. 99 Ω, 0.033 μF; C. 11 Ω, 0.3 μF; D. 11 Ω, 0.033 μF.

To obtain the total resistance of resistors in series we *add* the individual resistances to obtain, in this case, (33 + 33 + 33) Ω, i.e. 99 Ω.

To obtain the total capacitance of capacitors in series we first find its reciprocal by adding *the reciprocals* of the capacitances:

Reciprocal of total capacitance (μF) = $\frac{1}{0.1} + \frac{1}{0.1} + \frac{1}{0.1} = \frac{3}{0.1}$

and total capacitance is thus 0.1/3, i.e. 0.033 µF. Thus, the correct answer is 99 Ω, 0.033 µF which is answer B.

In answer A, the resistance is correct but the individual capacitances have been added to give the equivalent. This would be true only if the capacitors were in parallel.

In answer C the same error has occurred in calculating the equivalent capacitance and, in addition, the resistance is incorrect because it has been obtained using the reciprocal method, i.e. the method for parallel resistors.

In answer D the same error as in C has occurred in calculating the resistance. The capacitance answer is correct.

The method for *series* resistors and *parallel* capacitors is the same – *add* the individual values. The method for *parallel* resistors and *series* capacitors is the same – first find the reciprocal by adding the *reciprocals* of the individual values.

Specific objectives

The expected learning outcome is that the student:
2.14 States that a capacitor stores energy.
2.15 Defines energy stored by a capacitor as $\frac{1}{2}QV = \frac{1}{2}CV^2$.
2.16 Solves problems for energy stored in a capacitor.

Energy stored in a capacitor

When a voltage is applied to a capacitor a current flows in the capacitor leads for a certain time depositing charge upon the capacitor plates. We have seen that current is a movement of charged particles, the particles having been given energy by the voltage source (one volt is one joule per coulomb). Thus when the capacitor acquires an electric charge it also acquires energy. This energy is stored and may be used after the voltage supply has been removed.

What in fact happens is that an electric field is set up in the dielectric due to the charge and a p.d. thus exists between the capacitor plates, so that even after the voltage supply is disconnected a voltage remains across the capacitor.

If a conductive lead is now placed across the capacitor the energy of the capacitor is given up as an electric current flows in the lead and the voltage across the capacitor decays to zero as the charge moves off the plates and the electric field in the dielectric collapses.

The energy stored in a capacitor depends upon the charge held by the capacitor and the voltage and, since these quantities are related by the quantity capacitance, also on capacitance. In fact it can be shown that the energy stored by a capacitor $= \frac{1}{2} \times$ charge $\times$ voltage or, in symbols,

$$\text{energy} = \tfrac{1}{2}QV$$
$$\text{and since } Q = CV$$
$$\text{energy} = \tfrac{1}{2}CV^2$$

Either of these equations may be used depending upon which are the known quantities and which are the unknown.

Example 2.12 Calculate the energy stored by a 100 µF capacitor when 250 V is applied across it for a time sufficient for the capacitor to become fully charged.

$$\text{Energy} = \tfrac{1}{2} \times 100 \times 10^{-6} \times 250^2 \text{ (joules)}$$
$$= 3.125 \text{ J}$$

Example 2.13 Calculate the charge on a 0.5 μF capacitor which is storing 10 J of energy.

We are not given the voltage in this example but we do have two equations to help.

Using the usual symbols

$$Q = CV$$
$$\text{Energy} = \tfrac{1}{2}QV = \tfrac{1}{2}CV^2$$

From the energy equation: $10 = \tfrac{1}{2} \times 0.5 \times 10^{-6} \times V^2$

$$\text{so that } V^2 = \frac{10}{\tfrac{1}{2} \times 0.5 \times 10^{-6}}$$
$$= 4 \times 10^7$$
$$\text{and } V = 6.32 \text{ kV}$$

From the charge equation, $Q = 0.5 \times 10^{-6} \times 6.32 \times 10^3 = 3.16$ mC

Specific objective

The expected learning outcome is that the student:
2.17 *Lists and distinguishes between types of practical capacitor.*

Types of capacitor

Capacitors are usually classified by the material used for the dielectric. The more important characteristics of interest are the capacitance, working voltage, stability (change in capacitance during storage or when subjected to a changing environment), insulation resistance, tolerance (range of values which the capacitance may actually have for a given nominal value), physical size and price.

A summary of the main types available and notes on the characteristics are given below. The physical construction of capacitors varies and depends to some extent on the dielectric. Paper capacitors, for example, consist of interleaved layers of conductive foil and impregnated paper rolled up to form a cylinder; other capacitors may be flat, rectangular, square, disc or cylindrical. Various types and constructions are shown in fig. 2.5.

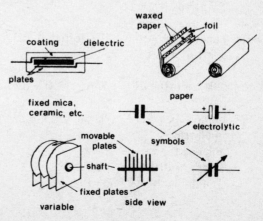

Figure 2.5

Variable capacitors used mainly in radio receiver circuits consist of a series of movable plates interleaved with a fixed set of plates. Movement of the movable plates changes the effective plate area of the capacitor and thus the capacitance.

Types of capacitor

Type	Construction and range of values	Working voltages	Stability	Tolerance	Use
Paper	Oil- or wax-impregnated paper 1 μF to 10 μF Bulky	Up to 600 V d.c.	Poor	± 20%	Power factor correction, motor, lighting circuits High-voltage smoothing, suppression circuits
Polyester	0.001 μF to 10 μF Compact	Up to 750 V d.c.	Good	± 20%	Not used at high frequency
Polystyrene	10 pF to 10 000 pF Small	Up to 160 V d.c.	Good	± $2\frac{1}{2}$%	
Polycarbonate	0.01 μF to 10 μF	Up to 630 V d.c.	Good	± 20%	Timing, high-stability applications
Mica	Silvered mica sheet 2 pF to 10 000 pF	Up to 350 V d.c.	Excellent	± 1%	Tuned circuits, filters, high-stability applications. Expensive
Ceramic	Metal-coated ceramic discs 2.2 pF to 0.1 μF	Up to 750 V d.c.	Excellent	± 2%	High-frequency circuits
Mixed dielectric	Paper/polyester 0.001 μF to 1 μF	Up to 1000 V d.c.	Good	± 20%	General purpose
Polypropylene	0.01 μF to 0.1 μF	Up to 1250 V d.c.	Good	± 20%	High frequency
Electrolytic	Large capacitance, small size, polarised 0.1 μF to 10 000 μF (Can be up to 100 000 μF)	Up to 450 V d.c.	Good	+ 100% − 25%	Unless specially designed must be connected correctly D.C. only. Used in power supplies and high-capacitance applications

Summary

Electric charge is a characteristic of a body such that it sets up a force between itself and other bodies. There are two kinds of electric charge called positive and negative. The unit of measurement of electric charge is the coulomb, symbol C. All materials contain atoms, which in turn, consist of negatively charged electrons and positively charged protons. Uncharged bodies contain equal numbers of electrons and protons and since they carry equal amounts of opposite charges the net result is no overall charge. When a body acquires charge it is because it gains or loses electrons, becoming negative or positive respectively as a result.

Bodies carrying a similar charge repel each other; bodies carrying a dissimilar charge attract each other. In addition charged bodies attract uncharged bodies.

A charged body sets up an electric field around it, in which the force due to the charge can be felt. The force acting on a body carrying one unit of charge when it is situated in an electric field is called the electric field strength of the field. Electric field strength is measured in newtons per coulomb and the quantity symbol is E.

Lines drawn in a diagram representing an electric field showing the direction of action of the force in the field are called lines of electric flux, or simply, electric flux. We say that one coulomb of flux is associated with one coulomb of charge. Electric flux density is the electric flux contained in one unit of area, the flux acting at right angles to the area. The unit of electric flux density is the coulomb per square metre. The quantity symbol is D.

The potential energy of a charged body which is placed in an electric field is called the electrical potential or, simply, potential of the body. It is measured in joules per coulomb or volts. The quantity symbol is V.

The difference in potential of a body placed at two different points in an electric field is called the potential difference or p.d. between the points. It, too, is measured in volts and has the quantity symbol V.

The potential difference between two points divided by the distance between the points is called the voltage gradient between the points. It is measured in volts per metre and is, in fact, equal to the electric field strength of the field between the points. The voltage gradient in a material at the point where it breaks down and conduction takes place is called the dielectric strength of the material.

An electronic component designed to store electric charge is called a capacitor. It consists of two (or more) plates which when connected to a voltage source acquire charge. The charge acquired for a particular applied voltage is directly proportional to the voltage. The constant of proportionality is called the capacitance of the capacitor.

Capacitance is the charge acquired per unit voltage applied. It is measured in coulombs per volt or farads. The quantity symbol is C and the unit symbol is F.

When a capacitor of capacitance C farads is connected to a voltage V volts the charge acquired Q coulombs is given by $Q = CV$, which is the capacitor equation.

Capacitors consist of two or more plates separated by a material called the dielectric. The capacitance of a capacitor depends upon the area of the plates, the distance between them and upon the material used for the dielectric.

For a capacitor having a plate area A square metres and plate separation d metres, the capacitance C is given by

$$C = \varepsilon \frac{A}{d}$$

where ε is called the absolute permittivity of the dielectric and may be defined as the capacitance of a capacitor having a plate area of one square metre and a plate separation of one metre. The unit of absolute permittivity is the farad per metre.

Absolute permittivity of a material is also the ratio between the electric flux density (D) in a material and the corresponding electric field strength or voltage gradient (E). The equation is $D = \varepsilon E$.

Absolute permittivity is the product of the relative permittivity or dielectric constant of the material, ε_r, and the permittivity of free space (a vacuum), ε_0. Relative permittivity of a particular material is the ratio of the capacitance of a capacitor with the material as a dielectric to the capacitance of the same capacitor having a vacuum as a dielectric. Relative permittivity is a ratio, a pure number, and has no units.

When capacitors are connected in series so that they acquire the same charge, regardless of the value of their capacitance, the reciprocal of the equivalent capacitance is the sum of the reciprocals of the individual capacitances. Using symbols C_E, C_1, C_2, C_3 etc. for equivalent and individual capacitances,

$$\frac{1}{C_E} = \frac{1}{C_1} + \frac{1}{C_2} + \frac{1}{C_3}, \text{ etc.}$$

When capacitors are connected in parallel so that the same p.d. is applied across each one the equivalent capacitance is the sum of the individual capacitances. Using the above symbols,

$$C_E = C_1 + C_2 + C_3, \text{ etc.}$$

The energy stored by a capacitor in terms of its charge Q coulombs, capacitance C farads and the voltage across it V volts, is given by

$$\text{Energy} = \tfrac{1}{2}CV^2 \text{ or } \tfrac{1}{2}QV \text{ joules}$$

EXERCISE 2   (Use $\varepsilon_0 = 8.85 \times 10^{-12}$ F/m)

1. Calculate the charge on a 5 μF capacitor when 100 V d.c. is applied to it.

2. Calculate the capacitance of a capacitor which has a charge of 20 μC when 50 V is applied to it.

3. Determine the voltage across a 0.5 μF capacitor holding 200 μC of charge.

4. A capacitor stores 100 μC of charge when 50 V is applied to it. Calculate the charge stored if the voltage is increased to 200 V.

5. The capacitance of a certain capacitor is 10 μF. If the dielectric remains unchanged but the plate area and separation are halved what would be the new capacitance?

6. Calculate the capacitance of a parallel plate capacitor of plate area 0.002 m^2 and dielectric thickness 0.002 m if the relative permittivity of the dielectric is 5.

7. Calculate the absolute permittivity of the dielectric of a 0.5 μF capacitor having a plate area of 0.0025 m^2 and dielectric thickness 0.5 mm.

8. Determine the field strength in a dielectric of thickness 2 mm when 400 V is applied. If the dielectric strength of the material is 10 MV/m, would the dielectric break down?

9. At breakdown, the voltage across a certain dielectric of thickness 1 cm is 500 kV. What is the dielectric strength of the material?

10. If the plate area of a capacitor is halved and its dielectric thickness is doubled, the dielectric being unchanged, the capacitance:
    A. is doubled;   B. is divided by four
    C. is quadrupled;   D. remains the same

11. Calculate the total equivalent capacitance of the circuits shown in fig. 2.6.

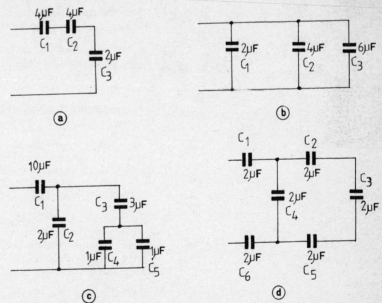

Figure 2.6

12. A capacitor having a dielectric constant of 2.5 has a plate area of 0.003 m^2 and dielectric thickness of 0.2 cm. A 550 V d.c. voltage is applied across the capacitor. Calculate the:
    (a) capacitance;
    (b) electric field strength in the dielectric;
    (c) charge stored by the capacitor;
    (d) energy stored by the capacitor.

13. Calculate the capacitance of a capacitor which stores 100 J of energy when 1000 V d.c. is applied. What is the charge on the capacitor?

14. The charge on a certain capacitor is 50 μC when 100 V is applied. The energy stored by the capacitor is:
   A. 2500 μJ;   B. 5 nC;   C. 0.5 μC;
   D. 2 MC.

15. Determine the charge on each capacitor in the circuits of fig. 2.6 when 100 V is applied across each circuit in turn.

16. Calculate the energy stored by each capacitor in the circuits shown in fig. 2.6 when 500 V d.c. is applied to the circuit as a whole.

SELF-ASSESSMENT EXERCISE 2

(Take $\varepsilon_0 = 8.85 \times 10^{-12}$ F/m)

Possible marks

1. State the units of electric charge, electrical potential and electric flux density. (3)

2. State the relationship between electric flux density D and electric field strength E. (3)

3. Define relative permittivity. (3)

4. Define capacitance. (3)

5. State the equation relating the capacitance of a capacitor C in terms of its plate area A, plate separation d and the dielectric permittivity. (3)

6. Give the equation for the equivalent capacitance C_E of capacitors C_1, C_2, C_3, etc. connected:
   (a) in series;
   (b) in parallel. (5)

7. Calculate the p.d. across a 0.1 μF capacitor when the energy stored in the dielectric is 50 mJ. (5)

8. The capacitance of a capacitor with air dielectric is 0.5 μF. If the dielectric is replaced by a material having a dielectric constant of 1000, all other factors being equal, what would be the capacitance?
   A. 500 μF;   B. 2000 μF;   C. 0.5 nF;
   D. It cannot be calculated without further details. (5)

9. Define relative and absolute permittivity. Calculate the absolute permittivity of the dielectric of a material of a 10 μF capacitor having a plate area of 25×10^{-4} m^2 and dielectric thickness 0.02 cm.
   If the dielectric is replaced by a material having a dielectric constant equal to twice that of the original what would be the new capacitance? (14)

10. What is meant by the dielectric strength of a material? The charge on a 10 μF capacitor when connected to a d.c. voltage supply is 100 mC. The dielectric strength of the capacitor dielectric is 1 MV/m. Calculate the dielectric thickness if the capacitor is on the verge of breakdown at this value of applied voltage. (14)

11. A capacitor having a dielectric constant of 3, a plate area of 4×10^{-3} m^2 and plate separation of 0.15 cm is connected to a 750 V d.c. supply. Calculate:
   (a) the capacitance;
   (b) the electric field strength;
   (c) the charge stored by the capacitor.
If this capacitor were connected to another identical capacitor such that each acquired the same amount of charge what would be the total capacitance? (14)

Capacitors and capacitance

12. Three capacitors of value $0.1\ \mu F$, $0.5\ \mu F$ and $1\ \mu F$ are connected in parallel, the combination being connected in series with a $0.4\ \mu F$ capacitor. The circuit is then connected to a 500 V d.c. supply. Calculate:
    (a) the charge on each capacitor;
    (b) the energy stored by each capacitor. (14)

13. In a certain capacitor connected to a 500 V d.c. supply the electric field strength in the dielectric is 500 kV/m, the flux density is $4.4 \times 10^{-4}\ C/m^2$ and the charge stored by the capacitor under these conditions is $4.4\ \mu C$. Calculate:
    (a) the capacitance;
    (b) the relative permittivity of the capacitor dielectric;
    (c) the plate area and plate separation. (14)

Answers

EXERCISE 2

1. $500\ \mu C$
2. $0.4\ \mu F$
3. 400 V
4. $400\ \mu C$
5. $10\ \mu F$
6. 44.25 pF
7. 10^{-8} F/m
8. 200 kV/m; No
9. 50 kV/m
10. B
11. (a) $1\ \mu F$ (b) $12\ \mu F$ (c) $2.42\ \mu F$ (d) $0.727\ \mu F$
12. (a) 33.19 pF (b) 225 kV/m (c) 1.83×10^{-8} C (d) $5\ \mu J$
13. $200\ \mu F$; 0.2 C
14. A
15. (a) $100\ \mu C$ on each.
    (b) $200\ \mu C\ (C_1)$; $400\ \mu C\ (C_2)$; $600\ \mu C\ (C_3)$.
    (c) $242.4\ \mu C\ (C_1)$; $151.5\ \mu C\ (C_2)$; $90.91\ \mu C\ (C_3)$; $45.46\ \mu C\ (C_4)$; $45.46\ \mu C\ (C_5)$.
    (d) $72.73\ \mu C\ (C_1)$; $18.18\ \mu C\ (C_2, C_3$ and $C_5)$; $54.5\ \mu C\ (C_4)$; $72.73\ \mu C\ (C_6)$.
16. (a) 31.25 mJ $(C_1$ and $C_2)$; 62.5 mJ (C_3).
    (b) 250 mJ (C_1); 1J (C_2); 2.25 J (C_3).
    (c) 73 mJ (C_1); 143 mJ (C_2); 34 mJ (C_3); 25.8 mJ $(C_4$ and $C_5)$.
    (d) 33 mJ (C_1); 2.065 mJ $(C_2, C_3$ and $C_5)$; 18.56 mJ (C_4); 33 mJ (C_6).

SELF-ASSESSMENT EXERCISE 2

Marks

1. Coulomb, volt, coulomb. (3)
2. $D = \varepsilon E$ where ε is the absolute permittivity of the material. (3)
3. Definition (see text). (3)
4. Definition (see text). (3)
5. $$C = \frac{\varepsilon A}{d}$$ (3)

6. (a) $$\frac{1}{C_E} = \frac{1}{C_1} + \frac{1}{C_2} + \frac{1}{C_3}, \text{etc.}$$ (2½)

(b) $C_E = C_1 + C_2 + C_3$, etc. (2½)

7. Energy $= \frac{1}{2} CV^2$
$50 \times 10^{-3} = 0.5 \times 0.1 \times 10^{-6} \times V^2$ (5)
from which $V = 1000$ V

8. A. (5)

9. Definition: absolute permittivity (3)
relative permittivity (3)

Using $C = \varepsilon \frac{A}{d}$

so that $\varepsilon = Cd/A = \dfrac{10 \times 10^{-6} \times 0.02 \times 10^{-2}}{25 \times 10^{-4}}$ (3)

$= 8 \times 10^{-7}$ (3)

The capacitance would double: 20 μF (2)

10. Definition: see text (3)

Using $Q = CV$ and thus $V = Q/C$

$V = (100 \times 10^{-3})/(10 \times 10^{-6}) = 10$ kV (4)

Dielectric strength = voltage gradient at breakdown

Thus $10^6 = 10 \times 10^3/d$, where d is the dielectric thickness (4)

and $d = 10^{-2}$ m (1 cm) (3)

11. (a) Capacitance $= \dfrac{\varepsilon A}{d} = \dfrac{8.85 \times 10^{-12} \times 3 \times 4 \times 10^{-3}}{0.15 \times 10^{-2}}$

$= 7.08 \times 10^{-11}$ or 70.8 pF (4)

(b) Electric field strength $= \dfrac{V}{d} = \dfrac{750}{0.15 \times 10^{-2}} = 500$ kV/m (4)

(c) Charge stored $= CV = 70.8 \times 10^{-12} \times 750 = 5.31 \times 10^{-8}$ C (4)

If the charge stored per capacitor is the same, the two are in series.

$$\frac{1}{C_E} = \frac{1}{70.8} + \frac{1}{70.8} \text{ (where } C_E \text{ is in pF)}$$

$C_E = 35.4$ pF (2)

12. (a) The circuit is equivalent to a 0.4 μF capacitor connected in series with a capacitor of capacitance $(0.1 + 0.5 + 1)$ μF, i.e. 1.6 μF. (1)

Equivalent circuit capacitance C_E (μF) is given by

$$\frac{1}{C_E} = \frac{1}{1.6} + \frac{1}{0.4} = \frac{2}{0.64}$$

hence $C_E = 0.32$ μF (1)

Charge taken from the supply Q (coulombs) is therefore

$Q = CV = 0.32 \times 500$ μC $= 160$ μC (1)

This charge is the charge on each series capacitor (or equivalent series capacitor) so that the charge on the 0.4 μF capacitor is 160 μC. (1)
The charge on the equivalent capacitor 1.6 μF is the same and practically is shared between the parallel-connected 0.1 μF, 0.5 μF and 1 μF capacitors according to the formula $Q = CV$. The voltage V must be known.

Voltage across parallel combination (using $V = Q/C$)

$$= \frac{160 \times 10^{-6}}{1.6 \times 10^{-6}} = 100 \text{ V} \tag{1}$$

Thus charge on:

$$0.1 \ \mu\text{F capacitor} = 0.1 \times 100 \times 10^{-6} = 10 \ \mu\text{C} \tag{1}$$

$$0.5 \ \mu\text{F capacitor} = 0.5 \times 100 \times 10^{-6} = 50 \ \mu\text{C} \tag{1}$$

$$1 \ \mu\text{F capacitor} = 1 \times 100 \times 10^{-6} = 100 \ \mu\text{C} \tag{1}$$

(b) Energy stored $= \frac{1}{2} CV^2$ (1)

For $0.4 \ \mu\text{F}$ capacitor, the voltage is $500 - 100$, i.e. 400 V. (1)

$$\text{Energy} = \frac{1}{2} \times 0.4 \times 10^{-6} \times 400^2 = 32 \text{ mJ} \tag{1}$$

For the other capacitors the voltage is 100 V (1)

Energy is:

$0.1 \ \mu\text{F} \quad \frac{1}{2} \times 0.1 \times 10^{-6} \times 100^2$, i.e. 0.5 mJ (1)

$0.5 \ \mu\text{F} \quad \frac{1}{2} \times 0.5 \times 10^{-6} \times 100^2$, i.e. 2.5 mJ (1)

$1 \ \mu\text{F} \quad \frac{1}{2} \times 1 \times 10^{-6} \times 100^2$, i.e. 5 mJ (1)

13. Supply voltage = 500 V
Electric field strength = 500 kV/m
Flux density = 4.4×10^{-4} C/m^2
Charge = 4.4 μC

(a) Capacitance $= Q/V = 4.4 \times 10^{-6}/500 = 8.8 \times 10^{-9} = 8.8$ nF (3)

(b) Permittivity $= D/E = \dfrac{4.4 \times 10^{-4}}{500 \times 10^3}$

$$\varepsilon = 8.8 \times 10^{-10} \text{ (F/m)} \tag{3}$$

Relative permittivity $\varepsilon_r = \dfrac{\varepsilon}{\varepsilon_0} = \dfrac{8.8 \times 10^{-10}}{8.85 \times 10^{-12}}$

$$\varepsilon_r = 99.44 \tag{2}$$

(c) Plate area = flux/flux density

$$= \frac{4.4 \times 10^{-6}}{4.4 \times 10^{-4}} = 10^{-2} \text{ m}^2$$

(The flux is 4.4 μC as is the charge) (3)

Plate separation $= \dfrac{\text{voltage}}{\text{electric field strength}}$

$$= \frac{500}{500 \times 10^3} = 1 \times 10^{-3} \text{ m} = 1 \text{ mm} \tag{3}$$

3 The magnetic field

Topic area: C

General objective — The expected learning outcome is that the student understands the laws relating to magnetic fields and applies them to series magnetic circuits.

Specific objectives — The expected learning outcome is that the student:
3.1 Defines the terms: flux, flux density, m.m.f. and magnetising force.
3.2 States the relationship between flux density B and field strength H.
3.3 Defines permeability as ratio of B to H.
3.4 Describes the effects of ferromagnetic materials on flux density.
3.5 Defines relative permeability.
3.7 States range of values of relative permeabilities for common ferromagnetic materials.
3.9 States the units of B, H, m.m.f., μ.

Magnetism has been known of for thousands of years and was used centuries ago as a simple means of navigating ships. Electro-magnetism, that is, magnetism due to electric current flow, was investigated in the latter part of the 19th century and originally was thought to be separate from magnetism shown by materials not connected in a circuit. We now realise that all magnetism is due to the movement of electric charge, since atomic theory tells us that all materials contain moving electrons.

The basic effects of magnetism have been considered in an earlier unit (Physical Science 1) and will not be re-examined. In this chapter we shall be looking more closely at magnetic circuits which have quantities similar in many ways to those in conductive circuits.

Basic quantities — In the region surrounding a magnet (or a conductor carrying current) in which a force due to magnetism may be felt, we say that the magnet (or conductor) sets up a *magnetic field*. If a tiny magnet is suspended freely in this field it sets in a certain direction depending upon its position in the field, and lines can be drawn on a diagram of the magnet and the surrounding area showing the various setting positions and thus the lines of action of the force set up by the magnet.

These lines are called *magnetic flux* and the diagram containing them is called the *magnetic field pattern*. Field patterns of bar magnets and conductors carrying currents are shown in fig. 3.1.

It should be realised that magnetic flux does not exist as a material thing; it is a device used in diagrams to show the magnetic force lines of action. The idea was suggested by Michael Faraday, and in earlier times the number of lines drawn per unit area was used to indicate how strong the effect of the magnetic field was; this is not done

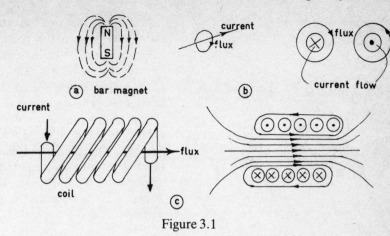

Figure 3.1

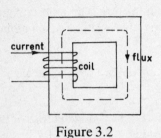

Figure 3.2

nowadays. Although magnetic flux is not a material quantity as such, we can nevertheless use the idea as a means of determining the effect of what is causing the magnetic field. To do this we consider a simple magnetic circuit as shown in fig. 3.2.

The simple magnetic circuit consists of a hollow square of suitable material (iron or one containing iron, for example) made up of four pieces called *limbs*. On one limb is wound a coil and when an electric current is passed through it a magnetic field is set up. It is found that the effect of the magnetic field cannot be felt outside the magnetic circuit so we know that the field is contained within the circuit limbs. This occurs because of the nature of the magnetic circuit material and does not happen if a non-magnetic material, for example, wood, glass, ceramic, copper or plastic is used. Only certain materials, particularly iron, iron compounds, nickel and cobalt display magnetic properties.

If the effects of the field are felt only within the magnetic circuit, flux lines may be drawn only within the circuit limbs and we thus arrive at the idea of a 'cause' and 'effect' and, in much the same way that in a conductive circuit the e.m.f. (the cause) sets up an electric current (the effect), we say that here a *magnetomotive force*, abbreviated m.m.f., sets up the flux. The m.m.f. is the cause and the flux is the effect.

The choice of name of the cause is unfortunate, as it is in the conductive circuit, because a magnetomotive force is *not* a force at all. As we have seen electromotive force is measured in units of energy per unit charge (joules per coulomb or volts) and, as we shall see shortly, magnetomotive force is not measured in force units. The names were chosen many years ago before the International System of Units came into being and we continue to use them because they are helpful in other ways.

(In a similar fashion we continue to show current in a conductive circuit flowing from the positive to the negative pole of a battery although we know that electrons, in fact, move the other way. We do this because a number of other helpful aids to memory have been devised which depend upon current flowing in a direction which was originally assumed to be correct.)

It is found that the strength of the magnetic field set up by the current flowing in the coil depends to a large extent upon the magnitude of the current and also the number of turns on the coil. Winding a conductor into the form of a coil 'adds together' the effect of the magnetic field due to each piece of the conductor and a much stronger field is established along the coil axis (see fig. 3.1).

The unit of m.m.f. is thus *defined* as the ampere-turn and to determine the m.m.f. for any coil of N turns carrying a current I amperes, the coil current and the number of turns are multiplied together to give IN ampere-turns. The abbreviation for the unit ampere-turn is 'At' or, since the size of the turn is unimportant and 'turns' are not measured in units, the abbreviation is sometimes written 'A' (as for current).

Magnetomotive force, m.m.f., is the ability to establish a magnetic field. It is measured in ampere-turns, abbreviated At or A. The symbol for m.m.f. is F.

In electrostatics we saw that electric flux is measured in terms of the electric charge associated with it, one coulomb of charge having one coulomb of flux. To find a way of comparing magnetic flux by measurement we turn to the voltage that is associated with magnetic flux when it changes. This is dealt with in more detail in chapter 4 under electromagnetic induction. For the moment we will accept a unit of flux, later it will be defined. The unit is, in fact, the *weber*, abbreviated Wb. The symbol for magnetic flux is ϕ (the Greek letter 'phi'.

Magnetic flux is a means of indicating the effect of a magnetic field. It is measured in webers, abbreviated Wb. The symbol for magnetic flux is ϕ.

In the simple magnetic circuit shown above the effect of the magnetic field is felt only within the circuit limbs and thus the magnetic flux is contained only within the limbs (assuming that the medium surrounding the circuit is totally non-magnetic). The cross-sectional area of the limbs is important for it is found that a smaller area tends to concentrate the effect of the magnetic field. To obtain a measure of this effect the magnetic flux is divided by the cross-sectional area of the limb, to give a quantity called *magnetic flux density*. Its units are webers per square metre (Wb/m^2), a special name for one weber per square metre being the *tesla*, abbreviation T. The symbol for magnetic flux density is B.

Magnetic flux density is a measure of the concentration of magnetic flux within a specified area. It is obtained by dividing the magnetic flux within the area by the area. The unit is the tesla (T), where one tesla = one weber per square metre, and the symbol is B.

In the electrostatic field we found that dividing voltage (the cause) by distance gave us a quantity equal to the electric field strength. Similarly in the magnetic circuit, dividing m.m.f. (the cause) by distance we obtain a quantity called magnetic field strength. It is measured in units of m.m.f. per unit of length, i.e. ampere-turns per metre (At/m or A/m) and the symbol is H.

Magnetic field strength is a measure of the effect of a magnetic field. It is obtained by dividing the m.m.f. causing the field by the distance over which the field acts. The unit is the ampere turn/metre (At/m or A/m), the symbol is H.

Magnetic materials

The ease with which a magnetic field is set up in a piece of any particular material depends upon the material itself. Not all materials are 'magnetic', that is they are not all easily magnetised or display magnetism. Those which are are grouped under the name *ferromagnetic* materials. To obtain some measure of the effect of the material we look again at the 'cause' and 'effect' within a *specified* size of the material. Within a particular size, that is of a given area and length, the 'cause' is measured in terms of the magnetic field strength – the m.m.f. per unit length. The 'effect' within the given area is the magnetic flux density – magnetic flux per unit area.

The ratio between the magnetic flux density, B, and the magnetic field strength, H, tells us how easy it is or otherwise to set up a particular value of flux density for a particular value of field strength. In some materials a far greater flux density will be set up by a particular value of field strength than in others. The ratio is called the *absolute permeability* of the material and is given the symbol μ.

To obtain the units of permeability we divide flux density B (Wb/m^2) by field strength H (A/m) to give webers per ampere-metre (Wb/Am). The relationship is written symbolically as $\mu = B/H$.

As is shown in the next chapter the weber/ampere is given the special name *henry* abbreviated H, so that the units of permeability are henry/metre (H/m).

The absolute permeability of a material is the ratio between magnetic flux density and the magnetic field strength in the material. It has the symbol μ and the units are henrys/metre (H/m) where one henry is one weber/ampere.

If measurements are made of magnetic quantities in a vacuum, the ratio of magnetic flux density to magnetic field strength so obtained is called the permeability of free space, symbol μ_0. The value of μ_0 in S.I. units is $4\pi \times 10^{-7}$ H/m.

Usually the absolute permeability of any material is written in terms of the permeability of free space using a quantity called *relative permeability*, symbol μ_r. Relative permeability is the ratio between the absolute permeability of a material and the permeability of free space. Symbolically,

$$\text{absolute permeability } \mu = \mu_r \mu_0 \text{ and } \mu_r = \frac{\mu}{\mu_0}$$

It is interesting to note at this point that the quantity permeability is very similar to permittivity in the electrostatic case. There, too, we had a relative permittivity ε_r, permittivity of free space ε_0 and absolute permittivity ε. The relationship defining permittivity is the same also, since

$$\text{Capacitance} = \frac{\varepsilon \times \text{area}}{\text{length}}$$

$$\text{and capacitance} = \frac{\text{charge}}{\text{voltage}}$$

$$\text{so that } \frac{\text{charge}}{\text{voltage}} = \varepsilon \frac{\text{area}}{\text{length}}$$

$$\text{and } \frac{\text{charge}}{\text{area}} = \varepsilon \frac{\text{voltage}}{\text{length}}$$

i.e. electric flux density = $\varepsilon \times$ electric field strength, which is similar to magnetic flux density = $\mu \times$ magnetic field strength.

The value of the relative permeability of a material tells us what kind of material it is magnetically. Values may range from about unity for non-magnetic materials (i.e. similar to a vacuum) to as high as 100 000 for magnetic materials. Materials in general may be classified according to their value of μ_r, as follows:

Diamagnetic materials have a value of μ_r less than 1. They include gold, silver, zinc, copper, bismuth, antimony and mercury. When subjected to a magnetic field they become very weakly magnetised.

Paramagnetic materials have a value of μ_r slightly greater than 1. They too may become weakly magnetised, slightly more so than diamagnetic materials but the difference is not significant. They include aluminium, platinum, oxygen, air, manganese and chromium.

For our purposes we may consider both diamagnetic and paramagnetic materials to be effectively non-magnetic.

Ferromagnetic materials become strongly magnetised when subjected to a magnetic field and may retain some or all of their magnetism when it is removed. Relative permeability for these materials varies considerably and some examples are:

    Pure annealed iron   200 to 5000
    Silicon iron   600 to 10 000
    Permalloy (78% nickel, 22% iron)   Up to 100 000
    Mu-metal (5% copper, 2% chromium, 75% nickel, 18% iron)   10 000 to 30 000
    Supermalloy (5% molybdenum, 79% nickel, 16% iron)
       Greater than 10^6

Note that all these alloys contain a percentage of iron, which is the principal magnetic material and gives rise to the general name *ferro*magnetic. The stem of this word (as in *ferrous* and *ferric*) is derived from the Latin name for iron.

Specific objectives

The expected learning outcome is that the student:
3.6   Draws comparative magnetising curves for typical ferromagnetic materials, e.g. cast iron, stalloy, a ferrite.
3.11 Lists the reasons for magnetic screening.
3.12 Defines hysteresis from given hysteresis loops.
3.13 Outlines the losses associated with hysteresis loops.

3.14 Identifies remanence, coercive force and saturation from hysteresis loops.

Magnetising curves

The value of relative permeability for a ferromagnetic material is not constant but varies according to the flux density within the material. This looks strange when it is first encountered since, as we have said, the value of permeability determines to some extent the flux density for a particular magnetic field strength. This is true but it is equally true the other way round in that the flux density affects the value of permeability. The reason is a phenomenon known as *saturation*.

It is believed that a ferromagnetic material is made up of a large number of 'molecular magnets' as shown diagrammatically in fig. 3.3. When unmagnetised these 'magnets' are distributed at random.

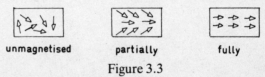

Figure 3.3

As the field strength is increased the 'magnets' begin to line up and the material becomes magnetised. Eventually the material becomes fully magnetised and is then said to be *saturated*. The flux density in the material rises as the field strength is increased, as shown in fig. 3.4 but the *rate of rise* (the slope of the B/H graph) becomes progressively less and eventually the curve levels out at a more or less constant value of flux density.

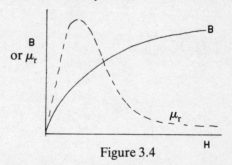

Figure 3.4

Permeability is equal to flux density divided by field strength and at any particular point on the B/H graph if the value of B at the point is divided by the value of H at the point, this gives the value of μ *at the point*. Doing this division at many points along the B/H graph gives a number of different values of μ and thus (by dividing by μ_0) of μ_r. Values of μ_r may then be plotted against values of H as shown in the figure.

As the B/H graph moves into saturation, B becomes relatively constant and H is increasing so that μ_r becomes progressively smaller. The reducing value tells us that B cannot be increased any further even though H may be.

Graphs of B against H for various materials are shown in fig. 3.5. Examination of the graphs shows clearly the relative values of flux density obtained for a particular value of field strength. The vertical line on the graph indicates a constant field strength shown as H_x and comparing the values of B, as given by reading the B axis corresponding to the points where the graphs cut the constant H_x line, show the relative ease of establishing magnetic flux in the materials. We can also see which material moves into saturation first as the value of H is progressively increased (see fig. 3.5).

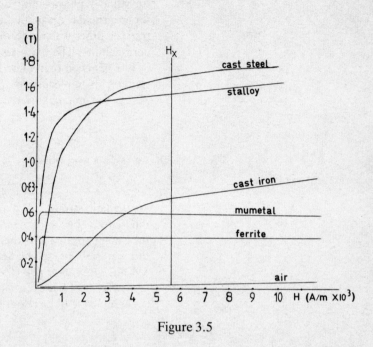

Figure 3.5

Hysteresis

When a ferromagnetic material is saturated the flux density reaches its maximum value. If now the field strength is progressively reduced, the flux density is also reduced but, for a particular value of field strength on the way down, the value of flux density is *higher* than it was at the value of field strength on the way up. The flux density *lags behind* as the field strength is made smaller. The phenomenon, which is due to an 'inertia' of movement of the molecular 'magnets' as they shift back to their random positions, is called *hysteresis* (from a Greek word meaning 'lagging').

When H reaches zero the material is not totally demagnetised and B has a value called the *remanent value*. Remanence is the retention of flux after the magnetising field strength has been reduced to zero. In order to reduce the value of B to zero, the magnetising field has to be *reversed* and eventually at a particular value of field strength in the opposite direction B becomes zero. This value of H is called the *coercive force* (although again, the word 'force' is misleading – it derives from the old name 'magnetic force' which was used for magnet field strength).

If H is now increased in this new (opposite) direction the material eventually saturates again, on this occasion behaving like a magnet which is oppositely polarised to the way in which it was when H was acting in the other direction. Again, if after saturation in the new direction, H is reduced, the values of B are not the same as they were for particular values of H on the way to saturation.

Again, when H once more reaches zero, B retains a value and the magnetising field must be reversed (back to the initial direction) and increased before B returns to zero.

The whole graph of B/H as this process is carried out is called a *hysteresis loop* and several kinds corresponding to different materials are shown in fig. 3.6.

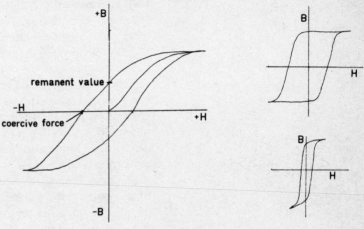

Figure 3.6

When a magnetic material is cycled through the loop, moving in and out of saturation in one direction then in opposite direction, a loss of energy occurs due to the internal movement of the 'molecular magnets' mentioned earlier. This energy is taken from the supply which provides the magnetising current (to set up the m.m.f.). It can be shown that the area contained inside the loop is directly proportional to the energy loss, so that a large area loop as in the figure indicates a material having much greater losses than a material having a low area loop.

If the magnetic material is part of a component to be used in a.c. circuits, e.g. transformers, chokes etc., each cycle of alternating current drives the material through the B/H loop and the energy loss occurs *in each cycle*. This can be quite serious so that usually a low-loss material is used in such applications. When the material is used in a.c. circuits it can be shown that the loss is directly proportional not to the frequency, as might be anticipated, but to the *square* of the frequency of the supply, so that it is more essential that a low-loss material is used in a.c. circuits.

Specific objectives

The expected learning outcome is that the student:
3.8   *Defines reluctance (S) as m.m.f./ϕ.*
3.9   *States the units of S.*
3.10 *Solves series magnetic circuits involving not more than a single change of dimension, material or airgap, using data from magnetisation curves.*

Magnetic circuit quantities and calculations

The quantities of interest in an electrically conductive circuit are e.m.f., p.d., current and resistance. Magnetic circuit quantities discussed so far include m.m.f., which is comparable to e.m.f., and magnetic flux, which is comparable to electric current. Quantities comparable to p.d. or resistance have not yet been discussed.

Potential difference is the energy used per unit charge in transferring charged particles from one point to another when conduction takes place. E.M.F. is the total energy per unit charge provided by a voltage source. The sum of the p.d.s in a conductive circuit is the total energy used per unit charge and equals the total energy provided per unit charge, the e.m.f. The p.d. between points in a conductive circuit can then be regarded as the proportion of the total e.m.f. used between those points.

The magnetic circuit quantity which is comparable to p.d. must then be a measure of the proportion of the total circuit m.m.f. used in a part of a magnetic circuit. To obtain it we look at electric field strength which between two points in a conductive circuit is measured by dividing p.d. by the distance between the points. If electric field strength is multiplied by the appropriate distance, the relevant p.d. is obtained.

Similarly, if magnetic field strength (H) is multiplied by the length of the part of the magnetic circuit in which it is measured, the quantity obtained is the magnetic circuit quantity comparable to p.d., that is, a measure of the proportion of the total m.m.f. required for that particular part of the magnetic circuit.

If all the magnetic field strength × length products are added together they equal the total m.m.f. This is written symbolically as

$$F = \Sigma H \ell$$

where F is the m.m.f. (At) and H is the magnetic field strength (At/m) across a length ℓ (metres) of the circuit.

The sign Σ is mathematical 'shorthand' meaning 'the sum of'. The $H\ell$ product does not have a special name but the idea is very useful in making magnetic circuit calculations. The magnetic circuit quantity comparable to resistance is called *reluctance*, symbol S.

To obtain resistance divide *voltage* by *current*. To obtain reluctance divide *m.m.f.* by *flux*, i.e.

$$R = \frac{E}{I} \text{ and } S = \frac{F}{\phi}$$

The unit is the ampere-turn per weber (At/Wb)

The reluctance of a magnetic circuit is the opposition to the setting up of magnetic flux. It is obtained by dividing the m.m.f. by flux. The symbol is S and the unit the ampere-turn per weber (At/Wb).

Comparable quantities

Conductive circuit
e.m.f., E, volts
current, I, amperes

Magnetic circuit
m.m.f., F, ampere-turns
flux, ϕ, webers

p.d., V, volts,   magnetic field strength × length, $H\ell$, ampere-turns

resistance, R, ohms.   reluctance, S, ampere-turns/weber

Reluctance of a magnetic circuit can be related to the dimensions of the circuit as follows:

Since
$$\text{flux density} = \frac{\text{flux}}{\text{area}}$$

and magnetic field strength $= \dfrac{\text{m.m.f.}}{\text{length}}$

and $\dfrac{\text{flux density}}{\text{magnetic field strength}} = \text{permeability},$

then $\dfrac{\text{flux}}{\text{area}} \times \dfrac{\text{length}}{\text{m.m.f.}} = \text{permeability}$

and by re-arranging, $\dfrac{\text{m.m.f.}}{\text{flux}} = \dfrac{\text{length}}{\text{permeability} \times \text{area}}$

But $\dfrac{\text{m.m.f.}}{\text{flux}} = \text{reluctance},$

so that $\text{reluctance} = \dfrac{\text{length}}{\text{permeability} \times \text{area}}$

or symbolically $S = \dfrac{\ell}{\mu A}$

where S, ℓ and A represent reluctance (A/m), length (m) and area (m²) respectively.

Sometimes a magnetic circuit is made up of a number of parts each of different reluctance. If the same flux links all the parts then the total reluctance is equal to the sum of the reluctances of the individual parts. This is comparable to resistors connected in series in a conductive circuit, in which the same current flows through all the resistors and the total resistance is the sum of the individual resistances.

The following magnetic circuit calculations should be studied carefully. A set of problems is at the end of the chapter.

Example 3.1 Calculate the flux density in a piece of iron of circular cross-section, radius 5 cm, if the flux in the iron is 0.5 Wb.

Area of cross-section $= \pi \times (5 \times 10^{-2})^2$

$= 78.54 \times 10^{-4} \text{ m}^2$

Flux density $= \text{flux/area} = \dfrac{0.5}{78.54} \times 10^4 = 63.66 \text{ Wb/m}^2$

The flux density is 63.66 T.

Example 3.2 A 200-turn coil is wound on a ring of magnetic material of radius 25 cm as shown in fig. 3.7. A current of 5 A is passed through the coil.

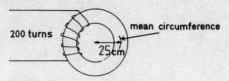

Figure 3.7

Calculate: (a) the m.m.f.; (b) the magnetic field strength.

$$\text{m.m.f.} = \text{coil turns} \times \text{current}$$
$$= 200 \times 5 = 1000 \text{ At}$$

The m.m.f. is 1000 At.

$$\text{Magnetic field strength} = \text{m.m.f.}/\text{length}$$

The 'length' in this case is the mean circumference of the ring which equals $2\pi \times$ radius, i.e. $2\pi \times 25 \times 10^{-2}$ m

$$\text{Magnetic field strength} = \frac{1000}{2\pi \times 25 \times 10^{-2}} = 636.62 \text{ At/m}$$

The magnetic field strength is 636.62 At/m.

Note that the magnetic circuit length is taken as the mean circumference of the ring, since because of the ring thickness there is an inner circumference and an outer circumference.

Example 3.3 The magnetic field strength in a magnetic circuit of length 45 cm is 2000 At/m. The flux density in the circuit is $1T$; the cross-sectional area of the circuit, which is uniform throughout, is 25 cm^2. Calculate: (a) the relative permeability of the material; (b) the m.m.f.; (c) the reluctance of the circuit.

(a) Absolute permeability μ = flux density/field strength and absolute permeability = relative permeability × permeability of free space, so that the relative permeability can be found from these two relationships.

$$\text{Flux density} = 1 \text{ T}; \text{ field strength} = 2000 \text{ At/m}$$

$$\text{Absolute permeability} = \frac{1}{2000} = 5 \times 10^{-4} \text{ H/m}$$

$$\text{and relative permeability} = \frac{5 \times 10^{-4}}{4\pi \times 10^{-7}} = 397.89$$

The relative permeability is 397.89

(b) $\quad\quad\quad\quad$ m.m.f. = field strength × length
$$= 2000 \times 45 \times 10^{-2} = 900 \text{ At}$$

The m.m.f. is 900 At.

(c) $\quad$ Reluctance = m.m.f./flux, and flux = flux density × area; so that flux = $1 \times 25 \times 10^{-4} = 25 \times 10^{-4}$ Wb.

$$\text{Reluctance} = \frac{900}{25 \times 10^{-4}} = 3.6 \times 10^5 \text{ At/Wb}$$

The reluctance is 3.6×10^5 At/Wb.

Example 3.4 An iron ring of mean length 85 cm and cross-sectional area 15 cm^2 is uniformly wound with 400 turns of wire of total resistance 250 Ω. The relative permeability of the material is 450. Calculate the reluctance of the magnetic circuit and the flux in the ring when 150 V d.c. is applied to the coil.

$$\text{Reluctance} = \text{length/absolute permeability} \times \text{area}$$

$$= \frac{85 \times 10^{-2}}{450 \times 4\pi \times 10^{-7} \times 15 \times 10^{-4}} = 10^6 \text{ At/Wb}$$

The reluctance is 10^6 At/Wb.

(Note that the absolute permeability $= 450 \times 4\pi \times 10^{-7}$ H/m)

$$\text{Flux} = \frac{\text{m.m.f.}}{\text{reluctance}} \text{ and m.m.f.} = \text{coil current} \times \text{coil turns}$$

The coil current $= 150/250 = 0.6$ A

m.m.f. $= 0.6 \times 400 = 240$ At

flux $= 240/10^6 = 240$ μWb

The flux is 240 μWb.

Example 3.5 An iron ring of mean length 25 cm and cross-sectional area 8 cm^2 has an airgap cut into it of width 2 mm. If the relative permeability of the iron is 900 calculate the total reluctance of the circuit.

Length of iron ring with airgap = 24.8 cm;
cross-sectional area = 8 cm^2

Hence $\quad \text{reluctance} = \dfrac{24.8 \times 10^{-2}}{900 \times 4\pi \times 10^{-7} \times 8 \times 10^{-4}} = 2.74 \times 10^5$

The airgap is of length 2 mm and has the same cross-sectional area as the ring.

Thus $\quad \text{reluctance of airgap} = \dfrac{2 \times 10^{-3}}{4\pi \times 10^{-7} \times 8 \times 10^{-4}}$

$$= 1.99 \times 10^6 = 19.9 \times 10^5$$

The total reluctance is then $(2.74 + 19.9) \times 10^5$, i.e. 22.64×10^5 A/Wb.

Note that the reluctance of the 2 mm airgap is over six times the reluctance of the 248 mm iron ring.

Example 3.6 The m.m.f. applied to a magnetic circuit is 1200 At. The circuit consists of a ring of magnetic material of mean length 50 cm excluding an airgap of width 5 mm which is cut into the ring. The flux density is found to be 0.05 T. Calculate the relative permeability of the iron.

Total m.m.f. = sum of $H\ell$ products

$$= (H_{iron} \times 50 \times 10^{-2}) + (H_{air} \times 5 \times 10^{-3})$$

where H_{iron}, H_{air} represent the magnetic field strengths in the iron and air respectively.

$$H_{air} = \frac{B}{\mu_0} = \frac{0.05}{4\pi \times 10^{-7}}$$

so that for the airgap,

$$H\ell_{air} = \frac{0.05 \times 5 \times 10^{-3}}{4\pi \times 10^{-7}} = 198.94 \text{ At}$$

and for the iron,

$$H_{iron} \times 50 \times 10^{-2} = 1200 - 198.94 = 1001.06$$

so that

$$H_{iron} = \frac{1001.06}{50 \times 10^{-2}} = 2002.12 \text{ At}$$

and

$$H_{iron} = \frac{B}{\mu_r \mu_0}, \text{ i.e. } 2002.12 = \frac{0.05}{\mu_r \times 4\pi \times 10^{-7}}$$

so that

$$\mu = \frac{0.05}{2002.12 \times 4\pi \times 10^{-7}} = 19.88$$

The relative permeability is 19.88

Example 3.7 A magnetic circuit has a mean length of 75 cm; 45 cm of this is of cross-sectional area 150 mm² and relative permeability 750, the remainder of the circuit being of cross-sectional area 275 mm² and relative permeability 1000.

Calculate; (a) the reluctance of the circuit; (b) the current required in a 250-turn coil to establish a flux density of 2 T in the higher permeability material.
Ignore flux leakage or fringing.

(a) $$\text{Reluctance} = \frac{\text{length}}{\mu \times \text{area}}$$

$$\text{Total reluctance} = \frac{45 \times 10^{-2}}{750 \times 4\pi \times 10^{-7} \times 150 \times 10^{-6}}$$

$$+ \frac{30 \times 10^{-2}}{1000 \times 4\pi \times 10^{-7} \times 275 \times 10^{-6}}$$

obtained by inserting the given values into the reluctance equation for each section of the circuit; note that the length of the higher permeability material is 75 − 45, i.e. 30 cm.

$$\text{Total reluctance} = 4.05 \times 10^6$$

(b) The cross-sectional area of the higher permeability material is 275 mm². The flux density is 2 T.

Thus the flux = flux density × area = $2 \times 275 \times 10^{-6} = 550 \ \mu\text{Wb}$.

If leakage and fringing are ignored this is the flux in the circuit as a whole.

Total m.m.f. = flux × reluctance = $550 \times 10^{-6} \times 4.05 \times 10^{6}$

$$\text{and current} = \frac{\text{m.m.f.}}{\text{coil turns}} = \frac{550 \times 4.05}{250} = 8.91 \text{ A}$$

Coil current is 8.91 A.

Example 3.8 If the length of a piece of magnetic material is doubled and its cross-sectional area is halved, the permeability remaining the same, the reluctance:

    A. remains the same;   B. is doubled;
    C. is halved;   D. is quadrupled.

Reluctance is directly proportional to length and indirectly proportional to area. If the length is doubled the reluctance is doubled, if the area is halved the reluctance is also doubled. The total effect is therefore that the reluctance is doubled twice or quadrupled. Answer D is therefore correct.

    Answer A assumes that length and area have the same effect so that doubling the length cancels out halving the area. For this to be true reluctance would have to be directly proportional to both length *and* area. Answer B ignores the effect of either length or area and answer C assumes the opposite effect to that which is true of either length or area while the effect of the other variable is ignored.

Example 3.9 The flux density in two pieces of identical magnetic material is the same. This implies that the two pieces of material:

    A. are of the same cross-sectional area;
    B. carry the same flux;
    C. are of the same length;
    D. have the same ratio of flux to cross-sectional area.

A. This is not so. The materials *may* have the same area, but flux density, which is the ratio flux:area, also depends on flux.

B. This is not so. Similarly, the flux *may* be the same in each piece but the flux density would not be the same unless the area of each piece were the same.

C. The two pieces *may* be of the same length, but unless the area is the same and the m.m.f. is the same (which is not stated) such that the flux is the same, the flux density will not be the same. Flux density is not directly related to length without involving other variables (m.m.f. and permeability).

D. This is the correct answer since flux density *is* the ratio of flux to cross-sectional area.

Example 3.10 If the magnetic field strength in a piece of magnetic material is doubled the flux density:

A. is doubled;   B. increases;   C. remains the same;
D. changes according to the state of magnetic saturation of the material.

Flux density and magnetic field strength are related by the magnetising curve (the B/H curve) of the material (see fig. 3.4). This curve is *not* linear throughout its range though it may be assumed to be approximately linear over a very small part of the range, i.e. for very small changes in magnetic field strength for example.

Answer A is not therefore necessarily correct. It MAY be approximately correct IF the change is very small and IF the material is not saturated.

Answer B may be correct and is, over most of the B/H curve, but if the material is fully saturated, flux density will not increase at all when magnetic field strength is increased.

Answer C may be correct if the material is totally saturated.

Answer D is the correct answer since the effect of magnetic field strength on flux density always depends on the state of magnetic saturation of the material.

Example 3.11   Fig. 3.8 shows a B/H curve for a magnetic material. A 1000-turn coil is wound on 15 cm of a piece of this material; determine the flux density in the material when the coil current is 262.5 mA. What is the relative permeability of the material at this value of flux density?

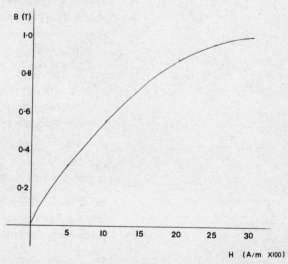

Figure 3.8

Magnetic field strength $H = \text{m.m.f.}/\text{length} = \dfrac{0.2625 \times 1000}{15 \times 10^{-2}}$

$= 1750 \text{ A/m}$

From the graph, when $H = 1750$ A/m, $B = 0.82$ T. The flux density is 0.82 tesla.

Absolute permeability $\mu = B/H = \dfrac{0.82}{1750}$ at this point

$$\mu = 4.686 \times 10^{-4}$$

and relative permeability $\mu_r = \mu/\mu_0$

$$\mu_r = \frac{4.686 \times 10^{-4}}{4\pi \times 10^{-7}} = 372.88$$

The relative permeability is 372.88

Example 3.12 A magnetic circuit of length 50 cm and cross-sectional area 5 cm^2 has a flux density of 0.6 T. The B/H curve is shown in fig. 3.8. Calculate the reluctance of the circuit and the current required in a 500-turn coil to establish this value of flux density.

$$\text{Reluctance} = \text{length/permeability} \times \text{area}$$

From the graph of fig. 3.8, when $B = 0.6$ T, $H = 1125$ A/m

$$\text{and permeability} = B/H = \frac{0.6}{1125} = 5.33 \times 10^{-4}$$

$$\text{Reluctance} = \frac{50 \times 10^{-2}}{5.33 \times 10^{-4} \times 5 \times 10^{-4}} = 1.876 \times 10^6 \text{ A/Wb}$$

Circuit reluctance is 1.876×10^6 A/Wb.

Magnetic field strength, $H = \text{m.m.f.}/\text{length}$

$H = 1125$ A/m; length $= 50$ cm

hence m.m.f. $= 1125 \times 50 \times 10^{-2}$

$$\text{Coil current} = \text{m.m.f.}/\text{turns} = \frac{1125 \times 50 \times 10^{-2}}{500} = 1.125 \text{ A}$$

Coil current is 1.125 A.

Summary

Magnetism is a characteristic of certain materials such that they exert a force on other materials. When this occurs the material exerting the force is said to be magnetised and is called a magnet. The region surrounding a magnetised material in which force is felt is called the magnetic field. Lines drawn to show the direction of action of the force are called lines of force, lines of flux or, simply, magnetic flux.

The ability to establish a magnetic field and magnetic flux is called magnetomotive force, abbreviated m.m.f., symbol F. It is measured in ampere-turns, At or A, and for a particular field is the product of the current and the number of turns of a coil in which the current would have to flow to establish such a field. The unit of magnetic flux is the weber, Wb, the quantity symbol being ϕ.

Magnetic flux density in a particular area is flux divided by the area, the flux acting at right-angles to the area. The symbol is B, the unit is the weber per square metre, called the tesla, symbol T.

Magnetic field strength, H, is the m.m.f. divided by the distance over which the m.m.f. acts. The unit is the ampere-turn per metre, At/m or A/m.

The absolute permeability of a material is the ratio between the flux density and the corresponding magnetic field strength. The symbol is μ and the units are henrys/metre, H/m.

The relationship between B, H and μ is $B = \mu H$. Absolute permeability is the product of relative permeability, μ_r, and the permeability of free space μ_0 (which equals $4\pi \times 10^{-7}$ H/m). Relative permeability is then the ratio between absolute permeability and the permeability of free space. The relationship is $\mu = \mu_r \mu_0$.

There are three kinds of material:
diamagnetic, in which μ_r is less than 1;
paramagnetic, in which μ_r is slightly greater than 1;
ferromagnetic, in which μ_r is many times greater than 1.

When magnetic flux density B is plotted against magnetic field strength H, the graph is not linear but flattens parallel to the H axis due to the material becoming saturated. Once this happens B cannot be increased further. The value of μ (which $= B/H$) thus rises and then falls to zero as H is increased.

Hysteresis is the 'lagging-behind' of flux density as magnetic field strength is altered. A full B/H graph with B and H being reversed in direction gives a characteristic loop called a hysteresis loop. Remanence is the value of B when H is reduced to zero (having been at saturation); the coercive force is the value of H required in the opposite direction to reduce B to zero. Hysteresis causes an energy loss in a magnetic material subjected to changing fields.

A magnetic circuit is an interconnection of magnetic materials carrying a magnetic flux and to which one or more m.m.f.s are applied.

The reluctance of a magnetic circuit or part of a circuit, symbol S, is the m.m.f. divided by the flux set up by the m.m.f. The unit of reluctance is the ampere-turn per weber, At/Wb or A/Wb. Its comparable quantity in the conductive circuit is resistance. The total reluctance of a number of parts carrying the same magnetic flux is the sum of the reluctances of the individual parts. A comparable situation in the electrically conductive circuit is the series connection of resistors.

EXERCISE 3

(Assume $\mu_0 = 4\pi \times 10^{-7}$ H/m)

1. Define the terms 'magnetic field strength' and 'magnetic flux density'. Calculate the magnetic field strength and magnetic flux density in a piece of magnetic material of length 25 cm, cross-sectional area 200 mm^2 and carrying a flux of 0.15 mWb, when an m.m.f. of 700 At is applied to the material.

2. What is meant by absolute and relative permeability of a magnetic material? A piece of magnetic material has a flux density of 0.17 T and a magnetic field strength of 2000 At/m. Calculate the absolute and relative permeability of the material.

3. Determine the relative permeability of the material in question 1.

4. A magnetic circuit of mean length 75 cm is uniformly wound with 550 turns of wire of total resistance 600 Ω. The relative permeability of the material is 1100. Calculate the flux density in the material when 150 V d.c. is applied to the coil.

5. Define 'reluctance' of a magnetic circuit. Calculate the reluctance of a piece of material of relative permeability 150, length 75 cm and cross-sectional area 275 mm².

6. If a total m.m.f. of 500 At were applied to the piece of material of question 5 calculate the (a) magnetic field strength; (b) magnetic flux density.

7. Calculate the length of a piece of material of relative permeability 1500 which has the same reluctance as an airgap of length 5 mm, both airgap and material having the same cross-sectional area.

8. An iron ring of mean length 50 cm has a 5 mm radial airgap cut into it. The area of cross-section is 375 mm². Calculate the flux density in the ring when an m.m.f. of 2500 At is applied to the circuit. The relative permeability of the iron is 850.

9. Two pieces of magnetic material have the same reluctance. The permeability of the one material is twice that of the other, the length of both pieces being the same, The ratio of the cross-sectional area of the piece having the lower value of permeability to that of the other is:
    A. 2;   B. 0.5;   C. 1;   D. 4.

10. The current flowing in a 750-turn coil uniformly wound onto a ring of magnetic material of relative permeability 400 is 4 A. If the ring were replaced by a ring of the same dimensions but of new material of relative permeability 800, for the flux density to remain the same the new m.m.f. would be:
    A. 3000 At;   B. 1500 At;   C. 6000 At;   D. 12 000 At.

11. An 80 cm length of material has a 1250-turn coil uniformly wound upon it. When a current of 1.44 A is passed through the coil the flux established in the circuit is 0.14 Wb. Calculate the cross-sectional area of the circuit if the magnetic material has the B/H curve shown in fig. 3.8.

12. A magnetic circuit of length 125 cm has a magnetic field strength of 1800 A/m and a flux of 0.15 Wb established within it. The material has the B/H curve shown in fig. 3.8. Calculate:
(a) the m.m.f. required; (b) the circuit cross-sectional area; (c) the relative permeability of the material for this value of magnetic field strength.

Possible marks

SELF-ASSESSMENT EXERCISE 3   (Take $\mu_0 = 4\pi \times 10^{-7}$ H/m)

1. Define absolute and relative permeability. (3)

2. An m.m.f. of 2000 A acts over a magnetic circuit length of 75 cm. Calculate the magnetic field strength. (3)

3. Calculate the cross-sectional area of a piece of magnetic material carrying a flux of 0.1 Wb and a flux density of 0.01 T. (3)

4. The flux density in a certain material is 0.35 T, the magnetic field strength is 2500 A/m. Calculate the absolute permeability. (3)

5. Calculate the relative permeability of a material having an absolute permeability of $4\pi \times 10^{-4}$ H/m. (3)

6. Calculate the reluctance of a piece of magnetic material of cross-sectional area 0.015 m², length 250 cm and relative permeability 750. (5)

7. Calculate the relative permeability of a piece of magnetic material in which the flux density is 0.1 T and the magnetic field strength 2000 A/m. (5)

8. Calculate the flux density in a magnetic circuit of relative permeability 1250 when the magnetic field strength is 1750 A/m. (5)

9. An iron ring of mean length 75 cm and relative permeability 450 has an airgap of 5 mm cut into it. The ring is uniformly wound with a 1200-turn coil and the cross-sectional area of the material is 20 cm². Calculate the reluctance of the circuit and the flux set up in the material when the coil current is 0.15 A. (14)

10. Define the terms magnetic flux density and magnetic field strength and give the relationship between them. Draw up a table of values of relative permeability against magnetic field strength for the material having the B/H curve of fig. 3.8. Take the following values of H: 500, 1000, 1500, 2000, 2500, 3000 A/m. Explain why the relative permeability varies as the magnetic field strength is varied. (14)

11. A magnetic circuit of two sections carrying the same flux of 0.6 mWb. One section is of length 25 cm, cross-sectional area 150 mm² and relative permeability 500; the other section is of length 50 cm, cross-sectional area 250 mm² and relative permeability 750.
Calculate: (a) the reluctance of the whole circuit; (b) the m.m.f. required for the whole circuit; (c) the flux density and magnetic field strength in each part of the circuit.
Neglect flux leakage. (14)

12. A coil is wound uniformly on an iron ring and a magnetic flux is established within the iron. When a radial airgap is cut into the ring, the coil current has to be increased to maintain the same flux. Explain why this is so.
An iron ring is made of the material having the B/H curve of fig. 3.8. The length of the ring is 50 cm before a 5 mm airgap is cut into the ring along its radius. Calculate the m.m.f. required to establish a flux density of 0.6 T in the circuit. (14)

13. (a) Compare the magnetic circuit quantities having the symbols F, S and ϕ with analogous quantities in the conductive circuit.
(b) Suggest laws for the magnetic circuit which are analogous to Kirchhoff's laws in the conductive circuit.
(c) Compare the magnetic quantity μ with the electrostatic quantity ε giving the units and any other relevant relationships in each case. (14)

Answers

EXERCISE 3

1. 2800 A/m; 0.75 T
2. 8.5×10^{-5} H/m; 67.64
3. 213.15
4. 0.253 T
5. 1.447×10^7 A/Wb
6. (a) 6.67×10^2 A/m (b) 0.126 T
7. 7.5 m
8. 0.563 T
9. A
10. B
11. 0.152 m²
12. (a) 2250 A (b) 0.163 m² (c) 406.7

The magnetic field 57

SELF-ASSESSMENT EXERCISE 3

Marks

1. Definition: see text (3)

2. Magnetic field strength = m.m.f./length
$$= \frac{2000}{0.75} = 2.67 \times 10^3 \text{ A/m} \quad (3)$$

3. Flux density = flux/area
so area = flux/flux density = $\frac{0.1}{0.01} = 10 \text{ m}^2$ (3)

4. Absolute permeability = $\frac{\text{magnetic flux density}}{\text{magnetic field strength}}$
$$= \frac{0.35}{2500} = 1.4 \times 10^{-4} \text{ H/m} \quad (3)$$

5. Relative permeability = $\frac{\text{absolute permeability}}{\text{permeability of free space}}$
$$= \frac{4\pi \times 10^{-4}}{4\pi \times 10^{-7}} = 1000 \quad (3)$$

6. Reluctance = length/(absolute permeability × area)
$$= \frac{250 \times 10^{-2}}{750 \times 4\pi \times 10^{-7} \times 0.015} = 1.768 \times 10^5 \text{ A/m} \quad (5)$$

7. Absolute permeability = $\frac{\text{magnetic flux density}}{\text{magnetic field strength}}$
$$= \frac{0.1}{2000} = 5 \times 10^{-5} \text{ H/m} \quad (2)$$

Relative permeability = absolute permeability/μ_0
$$= \frac{5 \times 10^{-5}}{4\pi \times 10^{-7}} = 39.79 \quad (3)$$

8. Flux density = absolute permeability × magnetic field strength
$$= (1250 \times 4\pi \times 10^{-7}) \times 1750 \text{ T} = 2.75 \text{ T} \quad (5)$$

9. Using $S = \frac{\ell}{\mu A}$ for each part of the circuit

For the iron
$\ell = 74.5$ cm (75 cm less airgap); $A = 20 \times 10^{-4}$ m^2; $\mu = 450 \times 4\pi \times 10^{-7}$

So $S = \frac{74.5 \times 10^{-2}}{450 \times 4\pi \times 10^{-7} \times 20 \times 10^{-4}} = 6.587 \times 10^5$ A/m (4)

For the airgap
$\ell = 5$ mm; $A = 20 \times 10^{-4}$ m^2; $\mu = 4\pi \times 10^{-7}$

So $S = \frac{5 \times 10^{-3}}{4\pi \times 10^{-7} \times 20 \times 10^{-4}} = 19.894 \times 10^5$ A/m (4)

Total reluctance = $(6.587 + 19.894) \times 10^5 = 26.48 \times 10^5$ A/m (2)

m.m.f. = $1200 \times 0.15 = 180$ A (2)

Flux = m.m.f./reluctance = $\frac{180}{26.48 \times 10^5} = 67.98$ μWb (2)

10. Definition: see text.
Relationship: $B = \mu H$, where μ is the absolute permeability of the material. (3)

Table of values:

B	0.32	0.55	0.75	0.88	0.96	1.00	(T)
H	500	1000	1500	2000	2500	3000	(A/m)
$\mu(= B/H)$	6.4	5.5	5	4.4	3.84	3.33	($\times 10^{-4}$ H/m)
$\mu_r (= \mu/\mu_0)$	509.3	437.7	397.9	350.1	305.6	265.3	–

(9) ($1\frac{1}{2}$ per value)

As the magnetic field strength is increased, the material approaches saturation and the value of flux density approaches a constant value. Thus B is becoming constant while H is increasing, μ (which is B/H) thus becomes smaller as H is increased towards saturation. (2)

11. Using $S = \dfrac{\ell}{\mu A}$

(a) Part one of circuit: Part two of circuit:
$\ell = 25$ cm; $\mu_r = 500$; $A = 150$ mm² $\ell = 50$ cm; $\mu_r = 750$; $A = 250$ mm²

$$S_1 = \frac{25 \times 10^{-2}}{500 \times 4\pi \times 10^{-7} \times 150 \times 10^{-6}} \quad S_2 = \frac{50 \times 10^{-2}}{750 \times 4\pi \times 10^{-7} \times 250 \times 10^{-6}}$$
$$= 26.53 \times 10^5 \qquad\qquad\qquad\qquad = 21.22 \times 10^5$$

(2 each)

Reluctance of whole circuit = $S_1 + S_2 = 47.75 \times 10^5$ A/m (2)

(b) M.M.F. = reluctance × flux = $47.75 \times 10^5 \times 0.6 \times 10^{-3}$ A = 2865 A (2)

(c) Flux density = flux/area

Part one of circuit: flux density = $\dfrac{0.6 \times 10^{-3}}{150 \times 10^{-6}} = 4$ T   ($1\frac{1}{2}$)

Part two of circuit: flux density = $\dfrac{0.6 \times 10^{-3}}{250 \times 10^{-6}} = 2.4$ T   ($1\frac{1}{2}$)

Magnetic field strength, $H = \dfrac{\text{flux density}}{\text{absolute permeability}}$

Part one of circuit: $H_1 = \dfrac{4}{500 \times 4\pi \times 10^{-7}} = 6366.2$ A/m   ($1\frac{1}{2}$)

Part two of circuit: $H_2 = \dfrac{2.4}{750 \times 4\pi \times 10^{-7}} = 2546.5$ A/m   ($1\frac{1}{2}$)

12. When a radial airgap is cut into an iron ring the reluctance of the magnetic circuit, now composed of iron and airgap, is considerably increased due to the much lower permeability of air compared with iron. To maintain the same flux before and after the airgap is cut, the m.m.f. will have to be increased and, since this depends directly on coil current, this, too must be increased. (2) (2) (1)

There are two parts to the circuit, the magnetic material and the airgap.

Flux density = 0.6 T.

Magnetic field strength, $H_{\text{mat}} = 1125$ A/m from fig. 3.8.

Length of material = 49.5 cm after airgap is cut

m.m.f. = $H_{\text{mat}} \times$ length = $1125 \times 49.5 \times 10^{-2} = 556.87$ A (3)

For the airgap: Magnetic field strength H_{air} = flux density/μ_0

$$= \frac{0.6}{4\pi \times 10^{-7}}$$

and m.m.f. = $H_{\text{air}} \times$ length = $\dfrac{0.6 \times 5 \times 10^{-3}}{4\pi \times 10^{-7}} = 2387.3$ A (3)

Total m.m.f. = $556.87 + 2387.3 = 2944.2$ A (3)

13. (a) *Magnetic circuit* *Conductive circuit*
F, m.m.f., ampere-turns E, e.m.f., volts (joules/coulomb)
S, reluctance, ampere-turns/weber R, resistance, ohms (volts/amperes)
ϕ, flux, webers I, current, amperes (coulombs/second)

(1 each) (6 total)

(b) *Kirchhoff's current law*

The algebraic sum of the currents at a junction is zero. An analogy for the magnetic circuit: the algebraic sum of the magnetic fluxes at a junction (of a magnetic circuit) is zero. (2)

Kirchhoff's voltage law

In any closed mesh the algebraic sum of the voltages is zero or the sum of the e.m.f.s is equal to the sum of the p.d.s.

For the magnetic circuit, the analogous quantity to p.d. is the product $H\ell$, which is the 'portion' of m.m.f. used over length ℓ. Accordingly, for the magnetic circuit: in any closed mesh of a magnetic circuit the sum of the m.m.f.s is equal to the sum of the $H\ell$ products in the circuit (where H represents the magnetic field strength over a length ℓ). (2)

(c) For the magnetic field: $\mu = \mu_r \mu_0$

where μ is the absolute permeability of the material, μ_r is the relative permeability and μ_0 ($4\pi \times 10^{-7}$ H/m) is the permeability of free space; permeability is the ratio of magnetic flux density (B) to magnetic field strength (H), i.e. $\mu = B/H$

For the electrostatic field: $\varepsilon = \varepsilon_r \varepsilon_0$

where ε is the absolute permittivity of the material, ε_r is the relative permittivity and ε_0 (8.85×10^{-12} F/m) is the permittivity of free space; permittivity is the ratio of electric flux density (D) to electric field strength (E), i.e. $\varepsilon = D/E$. (2)

4 Electromagnetic induction

Topic area: D

General objective The expected learning outcome is that the student understands how to: relate the law of electromagnetic induction to motor, generator and transformer principles.

Specific objectives The expected learning outcome is that the student:
4.1 Explains the motor principle in terms of the interaction between a magnetic field and a current-carrying conductor.
4.2 Explains the basis of the formula $F = B\ell i$, the linear relationship existing between F and other terms.

The main characteristic of magnetism is that a force is exerted by a magnetised magnetic material or magnet on other magnetic materials, whether or not they are magnetised. If these other materials are unmagnetised the force is one of attraction, the magnet attempting to pull the other material towards itself. If the other material is magnetised, i.e. is itself a magnet, the two magnets may attract or repel each other depending upon the direction of action of the magnetic fields in the region where they overlap.

This can easily be demonstrated using two bar magnets. Each magnet on its own will attract non-magnetised materials (provided they are ferromagnetic) such as iron filings, steel nails, tin cans, etc. When the two magnets are laid side by side and are pushed gently towards each other one of two things will happen. They may move together extremely easily and a definite force of attraction will be felt, or they will attempt to push each other apart and a force of repulsion will be felt. In this latter condition an effort must be made to hold the magnets together to counteract the force of repulsion.

A light bar magnet if freely suspended sets in a certain position due to the force exerted by the magnetic field of the earth (which itself acts like a very large magnet), and the end of the bar magnet pointing north is called the north-seeking or north pole of the magnet. The opposite end is called the south-seeking or south pole. When the two bar magnets are placed together the nature of the force is determined by the relative positions of these bar magnet poles. See fig. 4.1.

When like poles are adjacent the magnets repel each other. When unlike poles are adjacent they attract each other. The field patterns show that when the fields act in the same direction (a magnetic field is assumed to act from north to south pole) the force is repulsion. When they act in opposite directions the force is attraction.

The interaction of magnetic fields always occurs however the fields are produced, whether from a magnet, that is a piece of

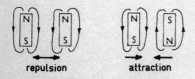

Figure 4.1

Electromagnetic induction

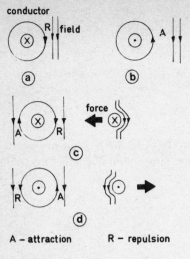

A – attraction R – repulsion

Figure 4.2

ferromagnetic material already magnetised, or by an electric current flowing in a conductor. The force produced in this way is used in the electric motor, in solenoids, relays, bells and door chimes. The direction of the force, attraction or repulsion, can always be determined by applying the basic principle, that like fields repel and unlike fields attract.

Consider a conductor carrying electric current, as shown in fig. 4.2, the conductor then being placed in a magnetic field due to something else (a magnet or another conductor carrying current).

The direction of the magnetic field due to an electric current is determined by the direction of flow of the current. It is easily remembered by the 'corkscrew rule' – a corkscrew is turned clockwise for the screw to move forward and anticlockwise for it to move backward; the direction of turn is the direction of action of the magnetic field produced by a current flowing (conventionally) in the direction of movement of the screw.

In fig. 4.2c the current is flowing into the page; the magnetic field is thus clockwise. If the other field is acting downwards as shown, on the left-hand side of the conductor we have unlike fields and attraction occurs. On the right-hand side of the conductor we have like fields and repulsion occurs. There is a force tending to move the conductor from right to left. The overall field pattern (fig. 4.2c) shows the distorted overall field and if the lines of force are thought of as elastic (trying to straighten themselves) the resultant force is indeed from right to left.

In fig. 4.2d the current is flowing out of the page, its field is reversed and if the other field stays the same, the force is from left to right. This method of determining the direction of action is the best one for it involves remembering first principles.

Force on a current-carrying conductor

The force exerted by a magnetic field on a conductor carrying current, due to the interaction of the field with that produced by the current, depends upon three things, the flux density of the main field, the conductor length and the size of the electric current.

The greater the flux density the greater is the force on the conductor. This is logical since the greater the flux density the more concentrated is the magnetic field per unit area. The force itself is due to interaction between fields so must depend on the magnetic field produced by the electric current. This field in turn depends on the electric current, so the greater the current, the greater the field produced and the greater the force between fields. Finally, the length of the conductor is important because the longer the conductor the larger is the region in which the conductor field and main field interact to produce the force.

Using B, ℓ and I to represent flux density (T), length (metres) and current (amperes) we can write:

$$\text{Force} \propto B\ell I$$

and the statement of proportionality can be made into an equation using a constant:

$$\text{Force} = \text{constant} \times B\ell I$$

If S.I. units are used, the force then being in newtons, the constant is unity and

$$\text{Force} = B\ell I$$

Example 4.1 A magnetic field has a flux of 0.5 Wb acting in an area of 2.5 cm^2. Calculate the force acting per metre length of a conductor carrying 4 A placed in the field.

$$\text{Flux density } B = \frac{0.5}{2.5 \times 10^{-2}} \text{ T}$$

Length $\ell = 1$ m and current $I = 4$ A.

$$\text{Hence force per metre length} = \frac{0.5}{2.5 \times 10^{-2}} \times 1 \times 4 \text{ N} = 80 \text{ N}.$$

Example 4.2 A magnetic circuit consists of an iron core of length 50 cm containing a 5 cm airgap. The core is uniformly wound with 2000 turns of wire carrying a current of 3 A. The cross-sectional area of the core is 15 cm^2 and the relative permeability of the core material is 580. When a separate current-carrying conductor is placed in the airgap it experiences a force of 1.5 N/m. Determine the current flowing in the separate conductor.

$$\text{Reluctance of a magnetic circuit} = \frac{\text{length}}{\text{permeability} \times \text{area}}$$

$$\text{Reluctance of iron} = \frac{45 \times 10^{-2}}{580 \times 4\pi \times 10^{-7} \times A} = \frac{6.17 \times 10^{-2}}{A}$$

where A (metre^2) represents the core area (note that the length of the iron is 45 cm).

$$\text{Reluctance of airgap} = \frac{5 \times 10^{-2}}{4\pi \times 10^{-7} \times A} = \frac{397.89 \times 10^2}{A}$$

Total reluctance $= (6.17 + 397.89) \times 10^2/A = 40406/A$

$$\text{Flux} = \frac{\text{m.m.f.}}{\text{reluctance}} = \frac{2000 \times 3}{40406} \times A$$

and since flux density $= \frac{\text{flux}}{\text{area}}$,

$$\text{Flux density} = \frac{\text{flux}}{A} = \frac{2000 \times 3}{40406} = 0.148 \text{ T}$$

Force on the current-carrying conductor is equal to flux density × length × current

so that force $= 1.5 = 0.148 \times 1 \times$ current

$$\text{and current} = \frac{1.5}{0.148}\text{A} = 10.1 \text{ A}.$$

Specific objectives

The expected learning outcome is that the student:
4.3 Explains the linear relationship existing between E and the other terms on the basis of the formula $E = B\ell v$.
4.4 Describes the production of an induced e.m.f. due to a changing magnetic field.
4.5 States Lenz's Law.
4.6 States Faraday's Law of electromagnetic induction.
4.7 Explains the generator principle in terms of Faraday's Law and Lenz's Law.

Electromagnetic induction

When an electric current flows in a wire a magnetic field is set up and if the field is allowed to interact with another magnetic field a force exists between the fields. If the conductor is free to move it will do so. There are three quantities here – *current* and a *magnetic field* producing *movement*.

Michael Faraday, when investigating these effects in the latter half of the nineteenth century, wondered if the situation was reversible as many are in nature. He discovered that it is and that if a conductor is moved through a magnetic field a current flows in the conductor, provided that it forms part of a closed circuit. Here, we have *movement* and a *magnetic field* producing *current*. The phenomenon is called *electromagnetic induction*.

Movement of the conductor is not in fact always necessary. What is necessary is that the magnetic field linking the conductor changes in some way as far as it affects the conductor. Thus moving a conductor through a magnetic field which is of constant value or changing the values of the field cutting a conductor has the same effect. An e.m.f. is *induced* across the conductor and if it is part of a closed circuit an electric current flows. Electromagnetic induction is the basic principle behind electrical generators (in which the conductors are moved) and transformers (where conductors remain stationary but the magnetic fields are changed in size).

Faraday discovered that the magnitude of the induced e.m.f. is directly proportional to the *rate of change of magnetic flux with time* or where a number of conductors are involved with a changing flux.

Induced e.m.f. $\propto$ number of conductors $\times$ rate of change of flux with time.

This is written symbolically as

$$E \propto N \frac{d\phi}{dt}$$

where E represents e.m.f., N represents the number of conductors and $\frac{d\phi}{dt}$ (pronounced dee phi by dee tee) is mathematical 'shorthand' for rate of change of magnetic flux (ϕ) with time (t). The branch of mathematics dealing with rates of change is called differential calculus and the use of 'd' as shown indicates a changing quantity.

If S.I. units are used the constant which is inserted to change this statement of proportionality into an equation is unity and the equation is

$$E = N\frac{d\phi}{dt}$$

where E is measured in volts, N is the number of turns and $d\phi/dt$ is measured in webers/second.

Another scientist, Lenz, discovered that an induced e.m.f. always acts *in a direction so as to oppose what is causing the induction*. In simple terms this means that, if the conductor across which the e.m.f. is induced forms part of a closed circuit and electric current flows, the current will flow in a direction such that its own magnetic field opposes the magnetic field causing the induction.

In fig. 4.3a a conductor is being moved from right to left against a magnetic field acting downwards. Induced current would flow in the conductor in a direction such that its magnetic field opposes the main field. This means that the flux lines must act downwards where the fields meet and thus in a clockwise direction round the conductor. The current must therefore be flowing in a direction *into* the paper. If current does not flow, the e.m.f. is still present and acts in a direction which would cause this direction of current flow if the current were able to flow. Because of the opposing nature of the induced e.m.f. it is sometimes called a *back e.m.f.* and a negative sign is inserted into the equation

$$E = -N\frac{d\phi}{dt}$$

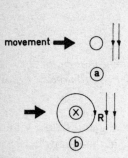

Figure 4.3

Example 4.3 A magnetic field of flux density 0.2 T acting over an area of 275 mm² is reversed at a constant rate of change over a period of 0.2 s. Calculate the induced e.m.f. in a coil having 150 turns which link the whole of the flux.

Flux = flux density × area = $0.2 \times 275 \times 10^{-4} = 55 \times 10^{-4}$ Wb.

Since the flux is reversed the *change* in flux is twice the value of flux:

Change in flux = $2 \times 55 \times 10^{-4}$ Wb, and this takes place in 0.2 s

so that

$$\text{rate of change} = \frac{2 \times 55 \times 10^{-4}}{0.2} = 55 \times 10^{-3} \text{ Wb/s}$$

Induced e.m.f. = number of turns × rate of change of flux

$$= 150 \times 55 \times 10^{-3} = 8.25 \text{ V}$$

Coil induced e.m.f. is 8.25 V.

Notice that the equation must use the instantaneous rate of change of flux, i.e. the rate of change of flux with time at any instant of time. In this case the average rate of change was calculated but

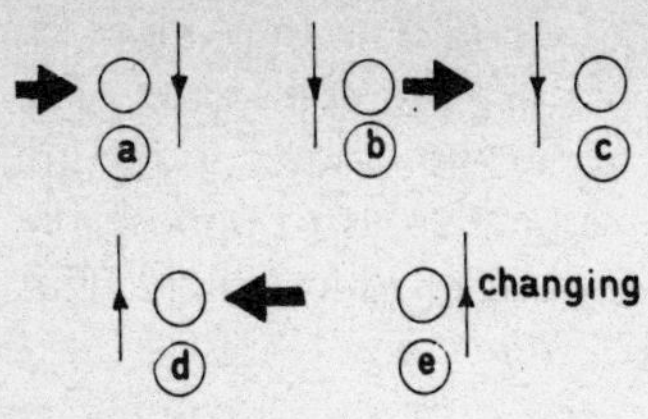

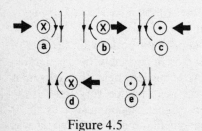

Figure 4.4

Figure 4.5

since it was stated that the rate of change was constant throughout the period the average rate of change may be taken as the instantaneous rate of change.

Example 4.4 Fig. 4.4 shows a conductor situated in a changing magnetic field. State the direction of current flow if the conductor forms part of a closed circuit.

To determine the direction of current flow, remember that the magnetic field set-up by the current opposes the cause of the electromagnetic induction. Draw in the lines of flux around the conductor so that this opposition occurs, then use the 'corkscrew rule' to determine in which direction the current must flow to set up this conductor field.

In fig. 4.5a, movement is from left to right and the magnetic field is acting downwards. Draw the conductor field in the same direction, i.e. acting clockwise, so that opposition occurs between the conductor field and the main field. To produce a clockwise conductor field, the current must flow *into* the paper.

In fig. 4.5b the conductor is being moved *away* from the field. The conductor field will act so that there will be attraction pulling the conductor back. Its flux lines are therefore clockwise and the current flow is again into the paper. Similar reasoning applied to figs 4.5c and 4.5d gives the current direction shown there.

In fig. 4.5e there is no movement of the conductor. The magnetic field is changing in size so that induction takes place. In this case if the field is acting upwards the conductor lines of flux are drawn upwards, i.e. anticlockwise, and the current flow is out of the paper. If the main magnetic field acts downwards, again the conductor flux is drawn downwards and the current flow would be into the paper. In both cases conductor field *opposes* the main field.

Specific objectives

The expected learning outcome is that the student:
4.8   Defines self-inductance and states its effects.
4.9   States the effect of self and mutual inductance.
4.10 Describes the transformer principle in terms of Lenz's law and the induced volts per turn.
4.11 Deduces for a transformer the effect of turns ratio on the voltage ratio.
4.12 Relates the equation $E = L\dfrac{di}{dt}$ to the equation $E = N\dfrac{d\phi}{dt}$ to show that $L \propto N^2$.
4.13 States the unit of inductance.
4.14 States that the energy stored by an inductor is $\tfrac{1}{2}LI^2$.
4.15 Solves problems relating to energy stored in an inductor.
3.11 Lists the reasons for magnetic screening.

Self-inductance

A changing magnetic field induces an e.m.f. across a conductor placed in the field. Whenever electric current flows it sets up its own magnetic field so that even if a separate field is absent, the conductor field, *if it changes*, induces an e.m.f. across the conductor. The conductor *self-induces* an e.m.f. and, as in the case of induction caused by a separate field, the e.m.f. will act so as to oppose the cause of the induction – in this case the existing conductor field. This characteristic of a conductor is called *self-inductance*.

Self-inductance means that whenever the current in a conductor changes its size (or direction), an e.m.f. is generated which acts so as to oppose the change. The e.m.f. in this case is called a *back e.m.f.* If the conductor current is being reduced (by reducing the applied voltage), the back e.m.f. opposes the reducing voltage and tries to maintain the level of current. If the current is being increased, the back e.m.f. again tries to prevent the increase. Whenever an electric circuit is switched 'on' or 'off' there will be a back e.m.f. trying to maintain the position before switching.

The effects of self-inductance depend particularly on the rate of change of the current with time and on the magnetic field generated by the current. All conductors have a magnetic field when current is flowing in them but the size of the field may be very small if the conductor is straight and in a non-magnetic medium such as air. If the conductor is coiled the magnetic field is much increased, as was explained earlier, and if the coil has a core of magnetic material the increase is even greater. Self-inductance has therefore much greater effects in coils, especially those with a magnetic material as a core.

When circuits containing coils and magnetic materials are opened the back e.m.f. can be so high that a spark occurs at the switch as the self-inductive effect tries to maintain current flow. This spark may be damaging so special switches or *suppressor circuits* may have to be used to reduce the spark. In certain applications, particularly the internal combustion engine in cars and other vehicles, the spark is used to start and maintain combustion. Here the circuit is designed to have maximum self-inductance.

When a circuit containing highly inductive components is closed current flow is not established immediately and the circuit takes time to settle down. The period of settling down is called the *transient* period and the initial voltages and currents are called transient voltages and currents (transient means 'changing'). The opposition to changing currents due to self-inductance is used in specially constructed inductive components called *inductors* or *chokes* (once known as 'choking coils') and these are employed to reduce or suppress undesired changes. Some examples are a.f. chokes, designed to reduce audio frequencies, and r.f. chokes, designed to reduce radio frequencies.

Self-inductance of a coil or conductor is defined using the back e.m.f. generated and the rate of change of current with time. The unit of self-inductance is the henry, symbol H (plural is *henrys* NOT henries). A conductor has a self-inductance of one henry when a current changing at the rate of one ampere per second self-induces a voltage of one volt. The defining equation is:

Back e.m.f. = self-inductance × rate of change of current with time

or, using symbols, $E = -L\dfrac{di}{dt}$

where E is the back e.m.f. (volts), L is the coefficient of self-inductance (henrys), and $\dfrac{di}{dt}$ is the rate of change of current with time (amperes per second).

Example 4.5 Calculate the self-inductance of a coil in which a current changing at the rate of 50 mA/s induces a back e.m.f. of 0.15 V.

Self-inductance $L = E/\dfrac{di}{dt}$ (the negative sign may be ignored unless direction of voltages is important).

so that $L = 0.15/(50 \times 10^{-3}) = 3$ H

Example 4.6 Calculate the back e.m.f. generated across a 15 H inductor when a current of 7 A flowing in it reverses its direction in 0.02 s.

The average rate of change of current is 14 A (twice 7 A since a reversal occurs) in 0.02 s, i.e. 14/0.02 A/s.

$$\text{Back e.m.f.} = \dfrac{15 \times 14}{0.02} \text{V} = 10\,500 = 10.5 \text{ kV}$$

As can be seen substantially high voltages can be self-induced.

Factors affecting self-inductance

There are two equations giving the value of induced e.m.f., one of which contains the coefficient of self-inductance. These equations may be used to give a third which relates self-inductance to other quantities.

The equations are $E = -N\dfrac{d\phi}{dt}$ and $E = -L\dfrac{di}{dt}$

Thus $L\dfrac{di}{dt} = N\dfrac{d\phi}{dt}$ and $L = N\dfrac{d\phi}{dt}\Big/\dfrac{di}{dt}$

Flux and current are related to each other by magnetic circuit quantities. Using the symbols:

B flux density
H magnetic field strength
ϕ flux
I current
A area of cross-section of magnetic circuit
ℓ length of magnetic circuit
μ permeability
N number of coil turns

$B = \mu H$, and since $B = \dfrac{\phi}{A}$ and $H = \dfrac{IN}{\ell}$,

$$\dfrac{\phi}{A} = \mu\dfrac{IN}{\ell} \text{ and } \phi = \dfrac{\mu A N}{\ell} \times I$$

i.e. Flux $= \dfrac{\mu A N}{\ell} \times$ current, so that

rate of change of flux with time $= \dfrac{\mu A N}{\ell} \times$ rate of change of current with time

$$\dfrac{d\phi}{dt} = \dfrac{\mu A N}{\ell} \times \dfrac{di}{dt} \text{ and } \dfrac{d\phi}{dt}\Big/\dfrac{di}{dt} = \dfrac{\mu A N}{\ell}$$

Substituting this in $L = N \dfrac{d\phi}{dt}\Big/\dfrac{di}{dt}$,

$$L = N \dfrac{\mu A N}{\ell} = N^2 \dfrac{\mu A}{\ell}$$

and it can be seen that self-inductance depends upon the square of the number of coil turns, on the permeability of the magnetic material used as a core and the dimensions of the magnetic circuit.

Example 4.7 A piece of magnetic material of length 10 cm and cross-sectional area 150 mm² is uniformly wound with a coil having 750 turns. The relative permeability of the material is 850. Calculate the self-inductance of the coil.

$$L = N^2 \dfrac{\mu A}{\ell}$$

$N = 750; \mu = 850 \times 4\pi \times 10^{-7}; A = 150 \times 10^{-6}; \ell = 10 \times 10^{-2}$

so that $L = \dfrac{750^2 \times 850 \times 4\pi \times 10^{-7} \times 150 \times 10^{-6}}{10 \times 10^{-2}} = 0.9 \text{ H}$

The coil self-inductance is 0.9 H.

Note that the equation $L = \dfrac{N^2 \mu A}{\ell}$

may be rearranged to give $\mu = \dfrac{L \ell}{N^2 A}$

The units of the right-hand side are $\dfrac{\text{henrys} \times \text{metres}}{\text{metres}^2}$

i.e. henrys/metre. (Number of turns is a pure number and has no units.)

Energy stored by an inductor

Energy is stored within the magnetic field of an inductor. It can be shown mathematically that the amount of stored energy is equal to half the product of the coefficient of self-inductance and the square of the current in the inductor.

Stored energy $= \tfrac{1}{2} L I^2$

This may be compared with the energy stored in the electric field of a capacitor (chapter 2) which is $\tfrac{1}{2} C V^2$.

Example 4.8 Calculate the energy stored in the magnetic field of a 10 H inductor in which a current of 2 A is flowing.

Energy $= \tfrac{1}{2} \times 10 \times 2^2$ joules $= 20$ J

Electromagnetic induction

Mutual inductance

The changing magnetic field which causes induction in a conductor may be produced by an electric current flowing in another conductor situated nearby. When this takes place it is said that *mutual inductance* exists between the coils. Mutual inductance is used to advantage in transformers, where, by adjustment of the number of turns on adjacent coils, voltage levels can be changed. This is discussed in more detail below.

Where the effects of mutual inductance are *not* required and coils are, of necessity, close to each other they may be placed at right angles to each other, which alters the relative directions of action of the coil fields, or they may be *screened* using a high permeability material which 'absorbs' the magnetic field, keeping its effects away from the coil being screened.

Mutual inductance between two coils is defined using the e.m.f. induced across one coil for a particular rate of change of current in the other coil. If a current changing at the rate of one ampere per second in one coil induces an e.m.f. of one volt across the other coil, the mutual inductance between the coils is one henry.

$$\text{Symbolically } E_1 = M \frac{di_2}{dt} \text{ or } E_2 = M \frac{di_1}{dt}$$

where E_1 is the induced e.m.f. across coil 1 when current is changing in coil 2

$\frac{di_2}{dt}$ is the rate of change with time of current in coil 2

E_2 is the induced e.m.f. across coil 2 when current is changing in coil 1

$\frac{di_1}{dt}$ is the rate of change with time of current in coil 1

and M is the coefficient of mutual inductance between the coils.

The units of E, $\frac{di}{dt}$ and M are volts, amperes/second and henrys respectively.

The transformer

The transformer consists of one or more coils mounted on a magnetic circuit so that all coils or parts of any one coil are linked by the same flux. Transformers are used *only* on alternating current supplies and the changing flux within the magnetic circuit core produces a changing induced e.m.f. across the coils.

Consider a two-coil transformer mounted on a core such as that shown in fig. 4.6. The left-hand coil has N_1 turns and a voltage V_1 is applied across it, causing a current I_1 to flow. This current sets up a magnetic flux, all of which (neglecting any loss or leakage) links with both coils. As a result of the changing flux a back e.m.f. shown as E_1 is induced across the left-hand coil and an e.m.f. shown as E_2 is induced across the right-hand coil.

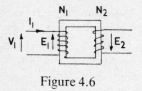

Figure 4.6

$$\text{Since } E_1 = -N_1 \frac{d\phi}{dt} \text{ and } E_2 = -N_2 \frac{d\phi}{dt}$$

(assuming that $\frac{d\phi}{dt}$, the rate of change of flux with time, is the same in both coils).

Then, by dividing one equation by the other,

$$\frac{E_1}{E_2} = \frac{N_1}{N_2} \text{ or } \frac{E_2}{E_1} = \frac{N_2}{N_1} \text{ and } E_2 = \frac{N_2}{N_1} E_1$$

The coil to which the voltage is applied is called the primary coil (or winding), the other coil is called the secondary coil (or winding).

Thus, secondary induced voltage = turns ratio × primary induced voltage

The flux in the core is set up by a small current flowing in the primary winding, called the *magnetising current*. The magnetising current is caused by the difference between the applied voltage V_1 and the primary induced voltage E_1 (which opposes V_1). Neglecting the very small magnetising current and the voltage causing it, the applied voltage V_1 and the induced primary voltage E_1 may be assumed equal and opposite.

The secondary winding induced e.m.f. may be used to supply a separate circuit, the supply voltage being different from the primary applied voltage, if this is desired, by arranging an appropriate turns ratio. A transformer in which the secondary voltage is greater than the primary voltage is called a step-up transformer. If the secondary voltage is less than the primary voltage, the transformer is called a step-down transformer.

Sometimes a 1:1 turns ratio is used and the primary and secondary voltages are equal; in this case the transformer merely electrically isolates the secondary circuit from the primary circuit and the transformer is called an *isolation* transformer. Isolation transformers are used for safety purposes in workshops and laboratories.

When a circuit is connected to the secondary winding, the transformer is said to be *loaded* and the secondary circuit is called the *load*. The induced secondary voltage sets up a current in the load which establishes a secondary winding m.m.f. Denoting this current by I_2, this m.m.f. would be equal to $I_2 N_2$, i.e. the secondary current × number of secondary turns.

When this happens an equal and opposite m.m.f. is established in the primary winding so that the magnetic circuit flux, in fact, remains unchanged. To set up the equal and opposite primary m.m.f. a current flows in the primary winding. Denoting this current by I_1, the primary m.m.f. set up when the transformer is loaded is equal to $I_1 N_1$. Since the primary and secondary m.m.f.s are equal (and opposite)

$$I_1 N_1 = I_2 N_2 \text{ and } I_2 = \frac{N_1}{N_2} I_1, \text{ whereas } E_2 = \frac{N_2}{N_1} E_1$$

and it is seen that the levels of current in the two windings are related by the reciprocal of the turns ratio which connects the levels of voltage in the windings, i.e. if the secondary voltage is greater than the primary voltage, the secondary current will be smaller and vice versa in the same proportion. The total primary current of a loaded transformer is actually the sum of the small magnetising current, which sets up the main flux, and the current I_1 (taking into account that they are not changing in the same way at the same time) but by neglecting the magnetising current, which is very small compared to I_1, we can say that

$$\frac{\text{primary current}}{\text{secondary current}} = \frac{\text{secondary turns}}{\text{primary turns}}$$

Similarly, neglecting the small difference between the primary applied voltage V_1 and its induced e.m.f. E_1 and the small difference between the secondary induced voltage and the actual voltage which appears at the secondary terminals on load we can also say that

$$\frac{\text{primary voltage}}{\text{secondary voltage}} = \frac{\text{primary turns}}{\text{secondary turns}}$$

(The actual secondary voltage is slightly less than the induced voltage due to the internal impedance of the secondary winding and the voltage drop across it when a load current is drawn).

A transformer is an extremely efficient component having efficiencies of up to 99%. Since the losses are so small there is little error in assuming that input power equals output power and, using V_2 to denote the secondary voltage applied to the load (slightly less than E_2)

$$V_1 I_1 = V_2 I_2 \text{ so that } \quad \frac{V_1}{V_2} = \frac{I_2}{I_1} = \frac{N_1}{N_2}$$

This approximation is very important and should be committed to memory.

Transformers are used widely in electrical and electronic applications to change voltage levels at high efficiency or for isolation purposes. Physical size ranges from very large oil-cooled transformers in steel tanks, seen at electricity generating stations, to very small low-power transformers used in electronic power supplies.

Example 4.8 A two-coil transformer has 1000 turns on the primary winding and 250 turns on the secondary winding. Calculate the secondary voltage when 250 V is applied to the transformer and the primary current when a secondary load current of 3 A is drawn from the transformer.

$$\text{Primary turns/secondary turns} = 1000/250 = 4$$

$$\text{Hence, secondary voltage} = \tfrac{1}{4} \times 250 = 62.5 \text{ V}$$

$$\text{and primary current} = \tfrac{1}{4} \times 3 = 0.75 \text{ A}$$

Example 4.9 A transformer having a primary/secondary turns ratio of 100 : 1 is connected to a resistive load. The applied voltage is 150 V and the primary current 20 mA. Calculate the power available from the secondary winding and the secondary voltage and current.

$$\text{Secondary voltage} = \frac{150}{100} = 1.5 \text{ V}$$

$$\text{Secondary current} = 100 \times 20 \times 10^{-3} = 2 \text{ A}$$

$$\text{Secondary power} = 1.5 \times 2 = 3 \text{ W}$$

The secondary power could be calculated directly from the primary quantities, assuming zero loss in the transformer:

$$\text{Primary power} = 150 \times 20 \times 10^{-3} = 3 \text{ W (as before)}$$

Summary

Magnetic fields acting in the same direction exert a force of repulsion on each other; magnetic fields acting in opposite directions exert a force of attraction on each other.

The force F newtons between a magnetic field of flux density B tesla and the magnetic field set up by a conductor of length ℓ metres situated in the magnetic field of density B and carrying a current I amperes is given by the equation

$$F = B\ell I$$

The e.m.f., E volts, induced in a coil of N turns situated in a magnetic field changing at the rate $\frac{d\phi}{dt}$ webers per second is given by

$$E = -N\frac{d\phi}{dt}$$

The minus sign indicates that the e.m.f. acts in a direction so as to oppose what is causing its induction. This usually means that if the e.m.f. is acting in a closed circuit a current would flow in a direction such as to set up a magnetic field acting in the opposite direction to the magnetic field causing the induction.

The e.m.f., E volts, induced across a conductor in which the current is changing at the rate of $\frac{di}{dt}$ amperes/second is given by

$$E = -L\frac{di}{dt}$$

where L is the coefficient of self-inductance and is measured in henrys, H. The conductor has a self-inductance of one henry if one volt is induced by a current changing at the rate of one ampere per second.

Since the induced e.m.f. E is equal to

$$N\frac{d\phi}{dt} \text{ and } L\frac{di}{dt},$$

$$\text{then } N\frac{d\phi}{dt} = L\frac{di}{dt} \text{ and } L = N\frac{d\phi}{dt}\Big/\frac{di}{dt}$$

The self-inductance L (H) of a coil of N turns wound on a core of absolute permeability μ (H/m), cross-sectional area A (m^2) and length ℓ (m) is given by

$$L = N^2 \mu A/\ell$$

The energy stored by an inductor of inductance L henrys and carrying a current of I amperes is $\frac{1}{2}LI^2$ joules.

When two coils are linked by the same changing flux set up by a current flowing in one of them, the e.m.f. induced across one coil E_1 volts is related to the rate of change of current with time, $\frac{di_2}{dt}$, amperes/second, in the other coil by the equation:

$$E_1 = M \frac{di_2}{dt}$$

where M is the coefficient of mutual inductance between the coils and is measured in henrys. If the voltage induced across one coil is one volt when the current in the other coil is changing at the rate of one ampere/second, the mutual inductance is one henry.

For a transformer, denoting primary and secondary e.m.f.s by E_1 and E_2 respectively, primary and secondary currents by I_1 and I_2 respectively and the number of turns on the primary and secondary windings by N_1 and N_2 respectively:

$$\frac{E_1}{E_2} = \frac{N_1}{N_2} \text{ and } \frac{I_1}{I_2} = \frac{N_2}{N_1}$$

These equations are approximately true for the primary and secondary p.d.s, also V_1 and V_2, since E_1 is approximately equal to V_1 and E_2 is approximately equal to V_2.

EXERCISE 4

1. Calculate the force acting on a conductor of length 15 cm situated in a magnetic field of flux density 0.32 T when a current of 3.5 A flows in the conductor.

2. The force per metre length of a conductor carrying 6 A situated in a magnetic field is 150 N. Determine the magnetic flux density of the field.

3. A single-turn coil is mounted between the pole pieces of a permanent magnet. The force per metre length acting on the coil when it carries a current of 150 mA is 7.5 N. The magnetic field flux is 1.25 Wb. Calculate the cross-sectional area of the pole pieces.

4. The reluctance of a magnetic circuit containing an airgap is 10^6 A/Wb, the circuit m.m.f. being 3500 A. The cross-sectional area of the airgap is 150 mm². Calculate the force acting per metre length on a conductor carrying 8.5 A placed in the airgap.

5. Calculate the average e.m.f. induced in a 200-turn coil when the magnetic flux linking the coil is changed uniformly from 250 mWb to 1 Wb in 0.2 S.

6. A magnetic field changing at the rate of 2 Wb/s links a coil causing an e.m.f. of 400 V to be induced across it. Determine the number of turns on the coil.

7. Fig. 4.4a shows a conductor being moved into a magnetic field. State the direction of:
    (a) Movement, if the current direction is out of the paper, the main field acting downwards.
    (b) Current flow, if movement is from left to right and the main field acts upwards.
    (c) The magnetic field, if movement is from right to left and current direction is into the paper.

8. Calculate the e.m.f. induced across a 5 H coil when a current of 4 A is reversed in 0.5 s.

9. The voltage induced across a 2 H coil is 250 V. Determine the rate of change with time of the coil current.

10. The rate of change of current in a coil having 200 turns is 4 A/s. The induced e.m.f. is 450 V. Determine:
    (a) The coil coefficient of self-inductance;
    (b) The rate of change with time of the flux linking the coil.

11. Calculate the self-inductance of a coil in which a current changing at the rate of 200 mA/s induces a back e.m.f. of 20 V.

12. Calculate the self-inductance of a coil having 1000 turns mounted on a piece of magnetic material of length 25 cm and cross-sectional area 200 mm^2 if the relative permeability of the magnetic material is 1250.

13. If the number of turns on the coil in question 12 is trebled, all other factors remaining the same, would the self-inductance:
    A. remain the same?   B. treble?
    C. be reduced to one third of its original value?
    D. be increased to nine times its original value?

14. Calculate the current flowing in a 10 H inductor when the energy stored in the magnetic field is 200 J.

15. Calculate the energy stored in a 4 H inductor carrying a current of 3 A.

16. The energy stored in an inductor carrying 2.5 A is 30 J. Calculate the average e.m.f. induced across the inductor when this current is reduced to zero in 0.05 s.

17. Two coils are mounted on a magnetic circuit such that the same flux links both coils. A current changing at the rate of 0.7 A/s in one coil induces an e.m.f. across the other coil of 100 V. Determine the coefficient of mutual inductance of the coils.

18. A transformer has three secondary windings having 1000 turns, 500 turns and 200 turns respectively. The primary winding has 400 turns. Calculate the secondary voltages available when 150 V is applied to the primary winding.

19. The primary voltage and current in a transformer on load are 200 V and 0.08 A respectively. Calculate the secondary voltage and current assuming a turns ratio of 5:1 (primary:secondary).

20. The primary and secondary voltages of a transformer are 250 V and 6.3 V respectively. Calculate the primary current when 100 mA is drawn from the secondary winding.

Possible marks

SELF-ASSESSMENT EXERCISE 4

1. Write down the equation relating the force F newtons which acts on a conductor of length ℓ metres carrying a current of I amperes and situated in a magnetic field of flux density B tesla. (3)

2. Calculate the force acting on a conductor of length 4 m carrying a current of 3 A and placed at right-angles to a magnetic field of flux density 0.6 T. (3)

3. Calculate the e.m.f. induced across a coil having 200 turns situated in a magnetic field charging at the rate of 0.2 Wb/s. (3)

4. Calculate the e.m.f. induced across a 10 H coil when the rate of change of coil current is 8 A/s. (3)

5. Calculate the secondary voltage in a transformer of turns ratio 2000:100 when 200 V is applied to the primary. (3)

6. The current in a coil is reversed from 10 A flowing in one direction to 10 A flowing in the opposite direction in 0.2 s. The average e.m.f. induced is 1000 V. Calculate the self-inductance of the coil. (5)

7. The force per metre length on a conductor carrying a current of 8 A and situated in a magnetic flux of 25 mWb between two magnetic pole pieces is 0.8 N. Calculate the area of cross-section of the pole pieces. (5)

8. The secondary voltage of a transformer is 20 V when 200 V is applied to the primary. The primary current is 0.5 A. Calculate the secondary current. (5)

9. An inductor consists of a coil wound on a magnetic core of relative permeability 750, cross-sectional area 8 cm^2 and length 20 cm. The e.m.f. induced across the coil when the rate of change of the coil current is 5 A/s is 250 V. Calculate (a) the number of turns on the coil; (b) the rate of change of flux with time when the current is changing at 5 A/s. (14)

10. Describe what is meant by the coefficients of self and mutual inductance.
   Two coils are mutually coupled so that the e.m.f. induced across one coil is 20 V when the current is changing at a certain rate in the other coil. The e.m.f. induced across the coil carrying the current is 10 V and its coefficient of self-inductance is 0.8 H. Calculate the coefficient of mutual inductance between the coils. (14)

11. Derive the equation relating the inductance of a coil L henrys to the number of turns on the coil N, the cross-sectional area of the coil core, a square metres, its absolute permeability μ, and its length ℓ metres.
   A 1000-turn coil is wound uniformly on a magnetic core of relative permeability 750, length 50 cm and cross-sectional area 1100 mm^2. Calculate the e.m.f. induced across the coil when the coil current is changing at the rate of 0.12 A/s. What is the rate of change of magnetic flux associated with this changing current? (14)

12. State the relationship between primary and secondary voltages and between primary and secondary currents in a transformer.
   The e.m.f. across a transformer secondary winding having 1000 turns is 125 V when 600 V is applied to the primary. Calculate:
   (a) the turns ratio of the transformer;
   (b) the number of turns on the primary winding;
   (c) the output power assuming zero loss when the primary current is 100 mA and the primary voltage is as given above;
   (d) the e.m.f. induced across the primary winding when the transformer core flux is changing at the rate of 0.1 Wb/s. (14)

13. (a) Using the equation relating reluctance S, of a magnetic core to the core length A, cross-sectional area ℓ, and permeability μ, and the equation relating the inductance of a coil L, to the number of coil turns N, and assuming the same core characteristics, obtain an equation relating S and L.
   (b) A magnetic core requires an m.m.f. of 2500 At to establish a flux of 25 mWb within it. Determine the number of turns on a coil wound uniformly on the core, if it has an inductance of 10 H. (14)

Answers

EXERCISE 4

1. 1.68×10^{-3} N
2. 25 T
3. 0.025 m^2
4. 198.3 N
5. 750 V
6. 200 turns
7. (a) Left to right
   (b) Out of the paper
   (c) Downwards
8. 80 V
9. 125 A/s

10. (a) 112.5 H (b) 2.25 Wb/s
11. 100 H
12. 1.2566 H
13. D
14. 6.32 A
15. 18 J
16. 480 V
17. 142.8 H
18. 375 V; 187.5 V; 75 V
19. 40 V; 0.4 A
20. 2.52 mA

SELF-ASSESSMENT EXERCISE 4

Marks

1. $F = B\ell I$ (3)
2. Force $= 0.6 \times 4 \times 3 = 7.2$ N (3)
3. Induced e.m.f. = turns × rate of change of flux
   $= 200 \times 0.2 = 40$ V (3)
4. Induced e.m.f. = self-inductance × rate of change of current
   $= 10 \times 8 = 80$ V (3)
5. Secondary voltage $= \dfrac{\text{secondary turns}}{\text{primary turns}} \times$ primary voltage

   $= \dfrac{100}{2000} \times 200 = 10$ V (3)

6. Using $E = -L\dfrac{di}{dt}$

   Average $\dfrac{di}{dt} = \dfrac{20}{0.2}$ (current is reversed) $= 100$ A/s.

   $E = -1000$ V, so $L = -E/\dfrac{di}{dt} = \dfrac{1000}{100} = 10$ H (5)

7. Force $F = B\ell I$, so $B = F/\ell I = \dfrac{0.8}{1 \times 8} = 0.1$ T (2)

   Flux density = flux/area

   so area = flux/flux density $= \dfrac{25 \times 10^{-3}}{0.1} = 0.25$ m² (3)

8. Using $\dfrac{V_1}{V_2} = \dfrac{I_2}{I_1}$ for a transformer where V and I represent voltage and current and subscripts 1 and 2 represent primary and secondary,

   $\dfrac{200}{20} = \dfrac{I_2}{0.5}$, hence $I_2 = \dfrac{0.5 \times 200}{20} = 5$ A (5)

9. (a) $E = -L\dfrac{di}{dt}$, hence $250 = L \times 5$ and $L = 50$ H. (3)

   Substitute this and the given figures into $L = \dfrac{\mu N^2 A}{\ell}$,

   $50 = \dfrac{750 \times 4\pi \times 10^{-7} \times 8 \times 10^{-4}}{20 \times 10^{-2}} N^2$ (2)

   Hence $N^2 = \dfrac{50 \times 20 \times 10^{-2}}{750 \times 8 \times 10^{-4} \times 4\pi \times 10^{-7}}$ and $N = 3642$ turns. (3)

   (b) Since $E = -N\dfrac{d\phi}{dt}$, (2)

   $\dfrac{d\phi}{dt} = \dfrac{E}{N} = \dfrac{250}{3642}$ Wb/s $= 68.64$ mWb/s. (4)

10. Description should include the equations (Description 2)

$E = -L\dfrac{di}{dt}$ and $E_1 = M\dfrac{di_2}{dt}$ and the definition of the henry. (Equations 2) (Definition 2)

The self-inductance of the coil carrying the current is 0.8 H; the induced e.m.f. is 10 V.

$$\text{Hence } 10 = 0.8\dfrac{di}{dt} \text{ and } \dfrac{di}{dt} = \dfrac{10}{0.8} = 12.5 \text{ A/s} \qquad (4)$$

The e.m.f. induced across the second coil is 20 V when the current in the first coil is changing at the rate of 12.5 A/s.

$$\text{Hence } 20 = M \times 12.5 \text{ and } M = \dfrac{20}{12.5} = 1.6\,\text{H} \qquad (4)$$

11. Derivation of $L = \dfrac{N^2 \mu A}{\ell}$ as text (5)

    Self-inductance of the coil,

    $$L = \dfrac{1000^2 \times 750 \times 4\pi \times 10^{-7} \times 1100 \times 10^{-6}}{50 \times 10^{-2}} = 2.07\,\text{H} \qquad (3)$$

    $$\text{Induced e.m.f.} = -L\dfrac{di}{dt} = -2.07 \times 0.12 = -0.248\,\text{V} \qquad (3)$$

    Since induced e.m.f. = −turns × rate of change of flux,
    rate of change of flux = −induced e.m.f./turns

    $$= \dfrac{0.248}{1000} \text{ Wb/s} = 0.248\,\text{mWb/s} \qquad (3)$$

12. $\dfrac{V_1}{V_2} = \dfrac{I_2}{I_1}$, where V_1, V_2, I_1, I_2 represent primary voltage, secondary voltage, primary current and secondary current respectively. (2)

(a) Turns ratio $\dfrac{\text{primary}}{\text{secondary}} = \dfrac{600}{125} = 4.8:1$ (3)

(b) $\dfrac{\text{Primary turns}}{\text{Secondary turns}} = \dfrac{4.8}{1}$

    Primary turns = 4.8 × secondary turns = 4.8 × 1000 = 4800 (3)

(c) Input power = primary voltage × primary current
    = 600 × 100 × 10^{-3} = 60 W (2)

    Output power = input power (if there are no losses) = 60 W (1)

(d) Induced e.m.f. = −turns × rate of change of flux = −1000 × 0.1
    = −100 V (3)

13. (a) $S = \dfrac{\ell}{\mu A}$ and $L = N^2 \dfrac{\mu A}{\ell}$ (2 each)

    From $S = \dfrac{\ell}{\mu A}$, $S\mu a = \ell$ and $\dfrac{A}{\ell} = \dfrac{1}{\mu S}$ (2)

    Substitute for (A/ℓ) in the equation for L,

    $$L = N^2 \mu \dfrac{A}{\ell} = N^2 \mu \left(\dfrac{1}{\mu S}\right) \text{ and } L = \dfrac{N^2}{S} \qquad (3)$$

(b) $S = \text{m.m.f.}/\text{flux} = \dfrac{2500}{25 \times 10^{-3}} = 10^5 \text{ A/Wb}$ (3)

    and since $L = 10$, using $L = \dfrac{N^2}{S}$

    $$N^2 = LS = 10 \times 10^5 = 10^6$$

    and $N = 1000$ turns (2)

5 Alternating voltages and currents

Topic area: E

General objective The expected learning outcome is that the student knows the concepts of alternating quantities.

Specific objectives The expected learning outcome is that the student:
5.1 Identifies alternating and unidirectional (sinusoidal and non-sinusoidal) waveforms from given sketches.
5.2 Defines the terms amplitude, period and frequency, and the values: instantaneous, peak-to-peak, r.m.s. and average.
5.3 Defines form factor.
5.4 Determines the approximate average and r.m.s. values of given sinusoidal and non-sinusoidal waveforms.
5.5 States that the average value, the r.m.s. value and the form factor of a sine wave are 0.637 maximum, 0.707 maximum and 1.11 maximum respectively.
5.6 Relates and calculates the quantities defined in 5.5 from given data.

Alternating voltages and currents

Electric current is a movement of electric charge, carried usually, but not always, by electrons. The direction of movement of the charge carriers is determined by the polarity of the applied voltage which is causing current flow. If the charge carriers are electrons the movement is from the negative side to the positive side of the supply, since electrons are negatively charged and like charges repel and unlike charges attract.

If the points of connection to the circuit of the applied voltage remain at the same polarity, the positive point remaining positive and the negative point remaining negative, the direction of current flow will also remain unchanged. A current having a direction of flow which also does not change is called a *unidirectional* or *direct* current, abbreviated to d.c. If, however, the polarity of the applied voltage peridiocally changes, the direction of current will change accordingly.

A current having a direction of flow which changes periodically is called an *alternating* current, abbreviated to a.c. The precise meaning of the word 'periodically' will be examined shortly. Similarly, the applied voltages which produce direct or alternating currents are called direct or alternating voltages respectively. Strictly speaking, one should not refer to a d.c. voltage or a.c. voltage since the 'c' in the abbreviation stands for current; it is, however, common practice.

When the direction of flow of an alternating current changes, its magnitude also changes since the current must fall to zero in the one

Alternating voltages and currents 79

direction before it rises to its new value in the other direction. Changing magnitude is therefore always associated with alternating currents (and, of course, voltages). The magnitude of a direct current may remain the same or may change; the word 'direct' tells us only that its direction of flow remains unchanged.

Fig. 5.1 shows a number of graphs in which are plotted current magnitude against time. The axes of the graphs are the same in each part of the figure. The vertical (or ordinate) axis is used to show current magnitude, the positive axis indicating a direction of current flow which is opposite to that indicated by the negative axis. The horizontal (or abscissa) axis indicates time.

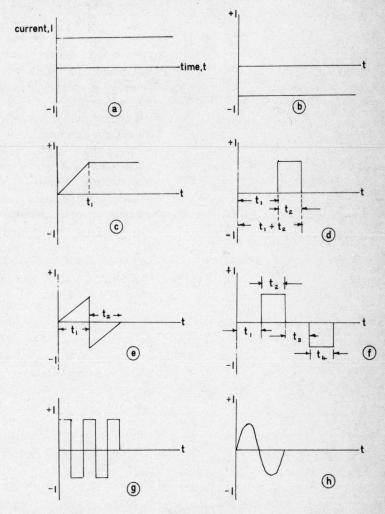

Figure 5.1

In part (a) of the figure the current shown is direct, since it does not change its direction, and is also of constant magnitude. Similarly, in part (b) the current is direct and of constant magnitude although on this occasion it is flowing in the opposite direction to the current shown in part (a). In parts (c) and (d) the current is direct but is of changing magnitude. The direct current shown in part (c) starts at zero, rises at a constant rate until after t_1 seconds its value stops changing and remains constant. The direct current shown in part (d) starts at zero, remains at zero for t_1 seconds then rises instantaneously to a value which is held constant for t_2 seconds. At the end of this period ($t_1 + t_2$ seconds from the start of timing) it drops instantaneously to zero. Its direction of flow does not change, however, so that it is a direct and not an alternating current.

The remaining four parts of the figure show alternating currents, the direction and magnitude of these currents changing as time goes on. In part (e) of the figure the current starts at zero, rises at a constant rate until after t_1 seconds it drops instantaneously to zero and then rises to a new value in the opposite direction. What happens now may be described in one of two ways. If we are considering magnitude alone, we can say that it falls from its new value and after a further t_2 seconds reaches zero. If we are considering magnitude and direction, then since the direction is opposite to that in which the current flowed when timing began and thus the magnitude is shown as a negative quantity, it can be said that the current *rises* from its negative value to zero, reaching zero after a further t_2 seconds. Sometimes the choice of word, 'rise' or 'fall', may cause confusion when diagrams such as these are explained (especially in descriptions of action of electronic circuits). This can be avoided by remembering the above point concerning whether we are considering magnitude alone or magnitude *and* direction.

In part (f) of the figure, the current starts at zero, remains at zero for t_1 seconds, rises instantaneously to a value at which it stays for t_2 seconds then falls instantaneously to zero for a further period of t_3 seconds before rising to a new value in the opposite direction. After a further t_4 seconds it falls again to zero.

In part (g) of the figure the current follows a similar process to that shown in part (f) except that on this occasion there are no periods of the current being at zero.

The current/time graph in part (h) of the figure is a special form known as a sinusoidal wave, sinusoid or sine wave. This special form is encountered very frequently since it is the form of the alternating current and voltage mains supply in the United Kingdom. A more detailed examination of this graph will be made shortly.

To summarise fig. 5.1, parts (a), (b), (c) and (d) show graphs of direct current, the currents in part (a) and (b) being of constant magnitude and those in parts (c) and (d) having varying magnitude. The rest of the figure shows graphs of alternating currents, the current in part (h) having a particularly important form since it is the one most commonly encountered.

Waveform definitions

Graphs of voltage or current against time are called *waveforms*. With any waveform of a regular recurring shape there are certain terms and definitions which we should know, so that waveforms of a similar shape may be compared. These terms are period, cycle, amplitude, frequency, instantaneous value and peak-to-peak, average and root-mean-square values.

Earlier the word 'periodically' was used. A regular waveform (as opposed to an irregular waveform) has a certain shape which recurs after a given time called the *period* of the waveform. Examine fig. 5.2.

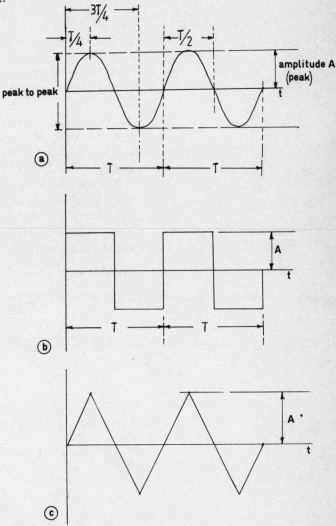

Figure 5.2

Part (a) shows a sine wave, part (b) a square wave and part (c) a triangular wave. In each case the wave follows a certain definite pattern which is repeated after T seconds, the period of each wave. The definite pattern which occurs during each period is called a *cycle* of the waveform.

In each case shown in fig. 5.2 the waveform rises to a maximum value in one direction and after a further time equal to half the period, rises to a maximum value in the opposite direction. This maximum value is called the *amplitude* of the waveform.

The number of cycles of a regular waveform which occur each second is called the *frequency* of the waveform. One cycle per second is called one *hertz*, which is abbreviated Hz. If each complete cycle lasts for a period of T seconds then the number of cycles in each second is equal to $1/T$ and, denoting frequency by f, we can write

$$f = \frac{1}{T}$$

which will be in hertz provided that T is measured in seconds. Multiples of the frequency unit include the kilohertz (kHz) which is 1000 Hz, the megahertz (MHz) which is one million hertz and the gigahertz (GHz) which is one thousand million hertz.

Example 5.1 The time taken for a sinusoidal voltage of amplitude 5 V to rise from zero to a peak positive value and then fall to −5 V is 0.3 s. Calculate (a) the periodic time of this waveform; (b) the frequency.

The amplitude or peak value of this voltage is 5 V. Consequently, when it acquires the value −5 V it has reached its negative peak and we are told that the time taken for it to rise from zero through a positive peak to the negative peak is 0.3 s. This time period must then be three quarters of the total periodic time (see fig. 5.2).

(a) The periodic time is thus

$$\frac{4}{3} \times 0.3 \text{ s, i.e. } 0.4 \text{ s or } 400 \text{ ms } (400 \times 10^{-3}\text{s})$$

(b) The frequency is equal to 1/periodic time

i.e. $\dfrac{1}{400 \times 10^{-3}}$ Hz, which equals $\dfrac{1000}{400}$ or 2.5 Hz.

Example 5.2 Determine the frequency of regular alternating waveforms having periodic times of: (a) 0.2 s; (b) 100 ms; (c) 125 μs.

In each case use $f = \dfrac{1}{T}$ where f is frequency (Hz) and T is periodic time (s).

(a) $T = 0.2$ s, thus $f = \dfrac{1}{0.2} = 5$ Hz

(b) $T = 100$ ms, i.e. 100×10^{-3} s

$$\text{thus } f = \frac{1}{100 \times 10^{-3}} = \frac{10^3}{100} = 10 \text{ Hz}$$

(c) $T = 125$ μs, i.e. 125×10^{-6} s

$$\text{thus } f = \frac{1}{125 \times 10^{-6}} = \frac{10^6}{125} = \frac{1000}{125} \times 10^3 = 8 \times 10^3 \text{ Hz} = 8 \text{ kHz}$$

Example 5.3 Determine the period of regular waveforms having the frequencies

(a) 50 Hz;   (b) 1 kHz;   (c) 5 MHz.

Since frequency = $\frac{1}{\text{period}}$, period = $\frac{1}{\text{frequency}}$

or $T = \frac{1}{f}$ using the symbols as before.

T is in seconds if f is in hertz.

(a) $f = 50$ Hz, thus $T = \frac{1}{50}$ s $= \frac{1000}{50}$ ms $= 20$ ms

(b) $f = 1$ kHz, i.e. 1000 Hz

thus $T = \frac{1}{1000}$ s $= \frac{1000}{1000}$ ms $= 1$ ms

(c) $f = 5$ MHz, i.e. 5×10^6 Hz

thus $T = \frac{1}{5 \times 10^6}$ s $= \frac{10^6}{5 \times 10^6} \mu$s $= \frac{1}{5} \mu$s $= 0.2 \mu$s

Notice the use of sub-multiples of seconds (milliseconds and microseconds) to obtain a neater answer. It is, of course, equally correct to write for (a) 0.02 s, for (b) 0.001 s, and for (c) 0.000 000 2 s but the use of a number of noughts following the decimal point can often lead to calculation errors. Multiples and sub-multiples of time and frequency units are always preferred where they reduce the possibility of error.

The magnitude of a non-varying quantity does not change and there can therefore be no confusion as to the value of the magnitude. A direct current 2 A, for example, means clearly that the current flows continually in one direction and has a constant value of 2 A. When the magnitude is changing and particularly when the direction of current flow is changing we must clearly define what is meant by the magnitude or value of the quantity. There are five such values commonly used – the instantaneous value, the peak-to-peak value, the peak value, the average value and the root-mean-square value.

Instantaneous value

The instantaneous value of a voltage or current which is varying in magnitude is the value at any given instant of time. If we wish to compare different voltages or currents at any particular instant then this value is useful. It is not however of much practical use if we wish to compare voltages or currents over a period of time, since the instantaneous value tells us nothing about how the voltages and currents are changing.

Peak-to-peak value

The peak-to-peak value, often abbreviated p/p, of an alternating voltage or current is the sum of the maximum value in one direction and the maximum value in the opposite direction. The maximum value in one direction is called the amplitude so that for a regular waveform equally displaced about zero, the peak-to-peak value is equal to twice the amplitude (see figure 5.2).

Average and root-mean-square values

An electric current may be used to provide heat and light, to drive motors and, in electronics, to convey information. In some of these uses it is the value of the current and how it varies over a period of time which is important; in others, as explained later, it is the value of the *square* of the current and how it varies over a period of time. The amplitude or peak value, the instantaneous value and the peak-to-peak value are all values which are not concerned with periods of time. The average value and the root-mean-square (r.m.s.) value are concerned with periods of time and their magnitude depends upon the time over which they are measured.

When an effect of an electric current depends directly upon its value (not the square of its value) and the electric current is changing in magnitude it is the average value of the current in which we are interested. The average value of a current or voltage which is changing in magnitude (whether it is alternating or direct) is the value of the constant magnitude direct current which would produce the same effect under the same conditions.

Effects such as heating and lighting involve the use of power or, over a period of time, energy. Power in an electric circuit depends directly upon the square of the current or voltage and here when the current or voltage is changing in magnitude it is the root-mean-square value which is most useful. The r.m.s. value of a current or voltage which is changing in magnitude (whether it is alternating or direct) is the value of the constant magnitude direct current which would produce the same power under the same conditions.

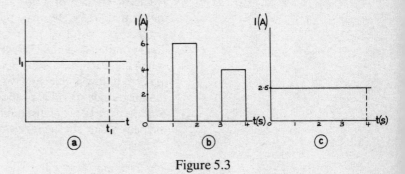

Figure 5.3

Examine fig. 5.3 part (a). It shows a graph of current plotted against time. The current rises from zero to a value I_1 at which it remains over the time period from 0 to t_1 seconds. The current is, of course, a direct current of constant magnitude. The average value here is the same as the maximum value, i.e. I_1, since the current does not change in magnitude.

Now examine part (b) of the figure. Here the current remains at zero for 1 second, then rises instantaneously to 6 A at which it remains for a further 1 second. At the end of this period the current falls instantaneously to zero. After a further period of 1 second the current again rises, this time to 4 A at which it remains for a final period of 1 second. The average value of this current depends upon

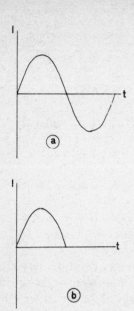

Figure 5.4

the period of time over which we require it. Over the first 1-second period the current is zero; over the first 2-second period the current is zero for half the time but has a value of 6 A over the other half of the period. Clearly, the average over the first 1-second period is different from that over the first 2-second period.

One method of determining the average value of a varying quantity over a particular period of time is by considering the *area* under the graph between the time limits. Consider again the graph in fig. 5.3 (b). The area under the graph between zero and 4 seconds consists of two rectangles, the first of height 6 (A) and width 1 (s), the second of height 4 (A) and width 1 (s). The total area is therefore 6×1 added to 4×1, i.e. 10. (The units are ampere-seconds but this need not concern us at present.)

The average value of this current may now be determined by dividing this area by the time over which we require the value. Thus over the 4-second period the average value is $10 \div 4$, i.e. 2.5 A. An instrument such as a permanent-magnet, moving-coil meter (which gives an indication determined by the average value) would respond over the 4-second period to the current shown in fig. 5.3 (b) in exactly the same way as to a current of constant magnitude of 2.5 A, the graph of which is shown in fig. 5.3 (c).

Fig. 5.4 (a) shows one cycle of a sinusoid. Over the period of the cycle the current rises to a peak value, falls eventually to zero, then rises to a peak value in the opposite direction, returning to zero in exactly the same way as it did in the positive direction. Over the periodic time, therefore, the current changes its value in one direction in exactly the same manner as in the other direction and over the same period of time (half a cycle).

An instrument of the type mentioned earlier would not respond to this wave since the pointer would be deflected first in one direction in a certain manner over a certain period of time and would then be deflected in the opposite direction in the same manner over a period of time equal to the first. The average value of the sine wave over a complete cycle is thus zero. For this reason the term 'the average value of a sine wave' usually refers to a half-cycle period as shown in part (b) of fig. 5.4. By determining the area under half a cycle by one or other of the methods in mathematics (mid-ordinate method, Simpson's Rule or integration) and dividing this area by the time of the half-cycle, it can be shown that for a current or voltage varying sinusoidally,

Average value of the current or voltage = 0.637 of the peak (maximum) value of the current or voltage.

Example 5.4 Find the average value of the currents shown in fig. 5.5 (a) and (b) over the time period indicated.

Figure 5.5 (a): The waveform here is triangular, the current rising linearly from zero to 5 A in 5 seconds, then falling in the same manner from 5 A to zero in the next 5 seconds. The method of

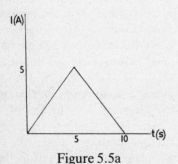

Figure 5.5a

finding the average value is to determine the area under the graph between 0 and 10 s and divide this by 10 s.

The area of a triangle is equal to $\frac{1}{2} \times$ base $\times$ height.

Base = 10; height = 5; area = $\frac{1}{2} \times 10 \times 5 = 25$

$$\text{Average value} = \frac{25}{10} = 2.5 \text{ A}.$$

This value may be deduced from the fact that the current is above the value 2.5 A for as long a period of time (from 2.5 s to 7.5 s) as it is below this value (from 0 to 2.5 s and from 7.5 s to 10 s).

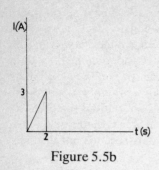

Figure 5.5b

Figure 5.5 (b): Area = $\frac{1}{2} \times 3 \times 2 = 3$; average value = $\frac{3}{2} = 1.5$ A

A similar comment applies here; the current is below 1.5 A from 0 to 1 s and above 1.5 A from 1 s to 2 s. An average value of 1.5 A is therefore to be expected.

The root-mean-square value of a current or voltage, as its name tells us, is the square root of the mean (or average) of the square of the current or voltage. To find the average of the square we can use the same technique as before, that is, divide the area under the graph over a certain period of time by the time. On this occasion, however, it is the graph of the *square* of the current or voltage. Having found the average of the square we then find its square root to give the r.m.s. value.

Consider again the current shown in fig. 5.3 (b). This is a direct current, since its direction does not change, but it is of changing magnitude. To find the equivalent constant-magnitude direct current we need to find the r.m.s. value of this current. The graph of the square of the current is shown in fig. 5.6. The graph consists of two one-second pulses of current, one rising to 36 A (6 × 6), the other to 16 A (4 × 4).

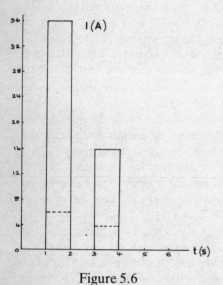

Figure 5.6

The area of the first pulse is 36 × 1 and of the second is 16 × 1. The total area over the 4 s period is therefore 36 + 16, i.e. 52. The mean of the square is therefore $\frac{52}{4}$, i.e. 13, and the root of the mean of the square is $\sqrt{13}$ which is 3.6 A.

Earlier we found the average value of this current to be 2.5 A, the r.m.s. value is 3.6 A. This means that to produce the same deflection on a permanent-magnet, moving-coil meter or to produce a similar motor effect, for example, as the varying current, a direct current of a constant magnitude of 2.5 A is required over the 4 s period. To produce the same power over this period, however, we need a direct current of a constant magnitude of 3.6 A.

It will be remembered that the average value of an alternating current or voltage which is symmetrical about zero, i.e. the 'positive' half is exactly the same shape as the 'negative' half, is zero since the effect of the 'positive' current is equal and opposite to the effect of the 'negative' current – when the effect depends on the value and not the square of the value of the current.

When the effect depends upon the square of the value, however,

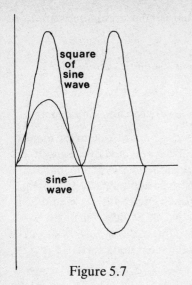

Figure 5.7

this is not so. The square of a negative quantity is positive and thus the graph of the square of a sinusoid will have the shape shown in fig. 5.7. In practical terms the power or energy dissipated by an electric current does not depend on the direction of current flow and therefore power is dissipated in both positive *and* negative half-cycles, the effect of the 'positive' current being equal but *not* opposite to the effect of the 'negative' current. It can be shown that the energy dissipated by an electric current over a period of time is proportional to the area under the current/time graph over that period of time.

By determining the area under the graph of the square of a current or voltage alternating *sinusoidally* (using one of the mathematical methods mentioned earlier), the mean of the square and then the root of the mean of the square, it is found that

$$\text{r.m.s. value} = 0.707 \times \text{peak value}$$

This can also be shown without using the more sophisticated mathematical methods as follows:

Figure 5.8 shows a sine wave over one cycle and the graph of the square of the sine wave. The sine wave is that of a current with a peak value I_M. The peak value of the graph of the square is thus I_M^2. A graph of a direct current of constant magnitude $\frac{1}{2}I_M^2$ is also drawn on the figure. Now to find the mean of the square we need to know the area beneath the curve over the period of the cycle. If these graphs are drawn properly to scale it will be found that the area A shown above the line of the direct current of value $\frac{1}{2}I_M^2$ is equal to the sum of the areas E and F below the line and area B above the line is equal to area G below the line. Areas C and D are common to both the area under the graph of the square of the current and the graph of the direct current.

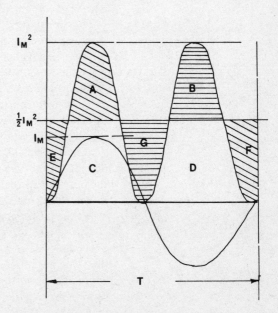

Figure 5.8

Area under the graph of the square of the alternating current

$$= A + B + C + D$$

Area under the graph of the direct current

$$= E + C + G + D + F$$

and since $A = E + F$ and $B = G$, the two areas are equal.

If we divide either area by the cycle period we obtain the mean or average over one cycle of the quantity represented by the graph. For the a.c. this is the mean of the square of the sine wave and for the d.c. this is the mean of a current of constant magnitude $\frac{1}{2}I_M^2$, i.e. is equal to $\frac{1}{2}I_M^2$ since the magnitude *is* constant. Since the areas under the graphs are the same the mean-of-the-square values will be the same. This tells us that the power delivered by the direct current would be the same as the power delivered by the alternating current under the same conditions.

Thus, mean of the square of the a.c. $= \frac{1}{2}I_M^2$ and

root of the mean of the square of the a.c. $= \sqrt{(\frac{1}{2}I_M^2)}$

$$= \frac{I_M}{\sqrt{2}} = 0.707\, I_M$$

and a direct current of constant magnitude of value $0.707\, I_M$ would produce the same power as the alternating current under the same conditions.

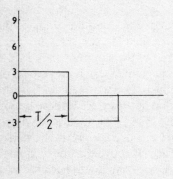

Figure 5.9

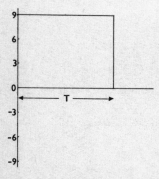

Figure 5.10

Example 5.5 Find the r.m.s. value of the current shown in fig. 5.9. Fig. 5.9 shows a square wave of peak value 3 A. The graph of the square of this current is shown in fig. 5.10.

The area under this graph over a cycle of the wave is $9 \times T$, where T is the periodic time, and the mean of the square is thus $9T \div T$, i.e. 9. The r.m.s. value is thus $\sqrt{9}$, i.e. 3 A and we see that for a square wave the r.m.s. value equals the peak value.

This is logical if one considers again the definition of the r.m.s. value. For the current in this example the value rises to 3 A in one direction and remains at this level for half a cycle. It then reverses instantaneously to a value of 3 A in the opposite direction for the next half-cycle. The only difference between a direct current of 3 A flowing for the periodic time and a cycle of this current is that for half the period, this current reverses. However, power is independent of direction of current flow and consequently this difference does not affect the power generated.

The square wave of peak value 3 A generates the same power over one cycle as a direct current of constant value 3 A flowing for the same period of time. The r.m.s. value of the square wave is the value of the constant magnitude direct current which generates the same power as the square wave and is therefore equal to 3 A, the peak value. This applies for all square waves regardless of the peak value.

Form factor

The form factor of a regular alternating waveform is the r.m.s. value divided by the average value.

For a sinusoidal wave of peak value I_M,

r.m.s. value = $0.707\,I_M$; average value = $0.637\,I_M$

$$\text{Form factor} = \frac{0.707\,I_M}{0.637\,I_M} = 1.11$$

For a square wave of peak value I_M,

r.m.s. value = I_M; average value = I_M

$$\text{Form factor} = \frac{I_M}{I_M} = 1$$

General objective

The expected learning outcome is that the student understands and uses phasor and algebraic representation of sinusoidal quantities.

Specific objectives

The expected learning outcome is that the student:
6.1 *Defines a phasor quantity.*
6.2 *Determines the resultant of the addition of two sinusoidal voltages by graphical and phasor representation.*
6.3 *Explains the phase-angle relationship between two alternating quantities.*
6.4 *Defines a sinusoidal voltage in the form $v = V_m \sin(\omega t + \phi)$.*
6.5 *Determines current from the application of a sinusoidal voltage to a resistive circuit.*
6.6 *Interrelates graphical, phasor and algebraic representation in the determination of amplitude, instantaneous value, frequency, period and phase of sinusoidal voltage and currents.*
6.7 *Determines power in an a.c. resistive circuit from given data.*

Graphical addition of alternating voltages and currents

When two or more direct voltages or currents of constant value are acting or flowing together in a circuit and it is desired to determine the resultant voltage or current, the addition is straightforward since the magnitudes of the quantities are not changing with time. Due regard must of course be paid to the direction of action of a voltage or of flow of a current whether the quantity is direct or alternating.

For example, consider the circuit of fig. 5.11. The total e.m.f. is due to 4 V acting from F to A and 2 V acting from A to F, i.e. producing a resultant e.m.f. of 2 V acting from F to A, and therefore circulating a clockwise current in mesh ABEF and mesh ACDF.

The total p.d. in mesh ABEF is the sum of 1 V across BE, with B positive with respect to E, and 1 V across EF, with E positive with respect to F, i.e. a total p.d. of 2 V.

The current flowing from A to B, of value 1 A, splits equally in the paths BE and CD to give two currents each of value 0.5 A. These currents recombine at E to give 1 A flowing from E to F.

We see that provided due care is taken over direction of action of voltages (e.m.f.s or p.d.s) or of flow of currents, the addition is straightforward, If, however, the e.m.f.s were alternating as in fig.

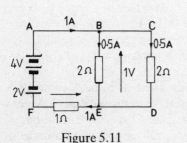

Figure 5.11

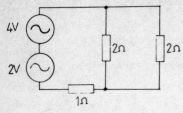

Figure 5.12

5.12 we would require more information concerning how the values of these voltages are changing with respect to each other and to time. We would require knowledge of their *phase relationship*.

Fig. 5.13 shows two alternating quantities shown as A and B, each part of the figure showing a different phase relationship between the two quantities.

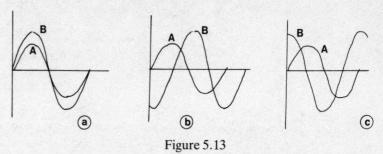

Figure 5.13

In part (a) of the figure, quantity A is rising and falling in exactly the same manner as quantity B and *at the same time*, i.e. both quantities reach their respective maximum values and reverse direction at the same time as each other. Here quantities A and B are said to be *in phase* with each other.

In parts (b) and (c) of the figure, quantity A reaches its peak and changes direction at a different time to quantity B. If the horizontal axis represents time, with time increasing as one moves from left to right, then quantity A reaches its peak in part (b) *before* quantity B and in part (c) *after* quantity B. In part (b) of the figure, quantity A is said to *lead* quantity B. In part (c), quantity A *lags* quantity B. Alternatively, in part (b) B lags A and in part (c) B leads A.

To determine the resultant voltage or current in a circuit containing a number of alternating voltages and currents we may examine the waveforms of the quantities and add them together, paying due regard to their direction at a number of points along the time axis, the greater the number of points the more accurate being the resultant waveform.

Example 5.6 Determine graphically the resultant current of those shown in fig. 5.14.

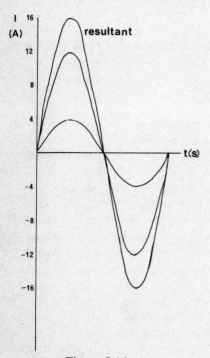

Figure 5.14a

Fig. 5.14 (a) shows two sinusoidal currents of peak value 12 A and 4 A respectively in phase with each other. The addition in this case is straightforward because the currents are in phase. The resultant is a sinusoidal current of peak value 16 A in phase with each of its two components.

Fig. 5.14 (b) shows two sinusoidal currents of peak value 10 A and 5 A respectively. The 10 A peak current leads the 5 A peak current by a quarter of a cycle. At t_0 the 10 A peak current has an instantaneous value of zero, the 5 A peak current an instantaneous value of -5 A. The resultant is thus -5 A. At t_1 the 10 A peak current has an instantaneous value of 10 A, the 5 A peak current an instantaneous value of zero. The resultant is thus 10 A. Similarly at t_2, t_3, t_4 the instantaneous value of the resultant is 5 A, -10 A and -5 A respectively.

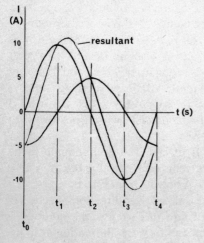

Figure 5.14b

The waveform of the resultant is as shown; careful drawing of the graphs on the correct paper will show it has a peak value of 11.18 A and leads the 5 A peak current by a time period less than one quarter of the period. The actual time will of course depend upon the period and thus the frequency. As was stated earlier, the greater the number of points taken the more accurate will be the waveform.

It will be noted that the examples each show two waveforms of the same frequency; the method can be used however to determine the resultant of any number of waveforms of different frequencies.

Graphical addition is an extremely accurate method of determining resultant waveforms and is especially useful if the components to be added are of differing waveshape (i.e. not all sinusoidal) or of different frequencies. The method is however laborious. An alternative method uses a single line to represent an alternating voltage or current, the length and position of the line relative to some fixed reference indicating the peak (or r.m.s.) value of the quantity and its phase relationship with other quantities. This line is called a *phasor*. Before looking further at methods of determining resultants using phasors we must now examine the basic theory behind them.

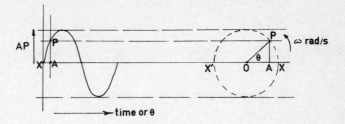

Figure 5.15

Phasors

A sinusoidally-varying quantity may be derived graphically by plotting a graph of the vertical distance from the horizontal of the end of a rotating arm against either time or angular displacement from the horizontal as the arm rotates (fig. 5.15). Thus, in fig. 5.15 the length AP varies sinusoidally as the arm OP rotates. As can be seen from the diagram,

$$AP = OP \sin \theta \text{ (}\theta \text{ is the angular displacement)}$$

and when θ is $\pi/2$ radians, AP = OP, which is the amplitude or peak value of the generated sine wave. Now if the arm is rotating with an angular velocity of ω rad/s the angle θ is given by

$$\theta = \omega t, \text{ where } t \text{ is time in seconds,}$$

and we can write

$$AP = OP \sin \omega t$$

For each revolution of the arm OP the angular displacement is 2π radians and one complete cycle is plotted. In each second, angular displacement is ω radians and since f cycles are plotted per second

$f = \frac{\omega}{2\pi}$, i.e. $\omega = 2\pi f$, so that AP = OP sin $2\pi ft$

Thus the horizontal axis of the sine wave may represent angular displacement ωt or time t as shown in fig. 5.15.

The length AP represents the instantaneous value of the sinusoidally alternating quantity, the length OP the maximum value, so that for a sinusoidally alternating current or voltage, the equations relating instantaneous values to the amplitude or peak values are

$$i = I_m \sin \omega t \text{ and } v = V_m \sin \omega t$$

respectively, where i, v represent instantaneous values, I_m, V_m represent peak values, and ω is equal to $2\pi \times$ frequency and is measured in radians per second. The time t or angle ωt is measured from the point at which the sine wave begins its positive excursion as shown.

If three arms OP_1, OP_2, OP_3 situated at fixed angles ϕ_1 between OP_1 and OP_2, and ϕ_2 between OP_2 and OP_3, are rotated at an angular velocity ω, as shown in fig. 5.16, the vertical displacements from the horizontal of the points P_1, P_2 and P_3, i.e. A_1P_1, A_2P_2 and A_3P_3 will plot three sine waves as shown.

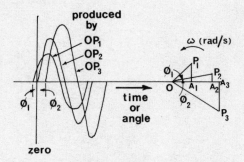

Figure 5.16

Examination of these sine waves shows that they pass through the same stage or phase at different times or angles. Thus the sine wave produced by OP_2 passes through zero going positive ϕ_1 radians after the sine wave produced by OP_1 has passed through this phase, and ϕ_2 radians before the sine wave produced by OP_3 passes through this phase. If the time when the OP_2 sine wave passes through zero going positive, is taken as $t = 0$, i.e. timing begins at this point, then the equation of this wave is

$$A_2P_2 = OP_2 \sin \omega t$$

as described above. The equations of the OP_1 and OP_3 waves are, respectively,

$$A_1P_1 = OP_1 \sin (\omega t + \phi_1) \text{ and } A_3P_3 = OP_3 \sin (\omega t - \phi_2)$$

These equations may be verified by plotting values of A_1P_1, A_2P_2, and A_3P_3 against t or ωt for any fixed value of ω. A brief examination at the point where $t = 0$, i.e. where $\omega t = 0$, shows that $A_2P_2 = 0$ (as in fig. 5.16), $A_1P_1 = OP_1 \sin \phi_1$ and $A_3P_3 = OP_3 \sin(-\phi_2)$ which is verified by the diagram. Also, A_2P_2 is zero when $\sin \omega t = 0$, i.e. when $\omega t = 0, \pi, 2\pi, 3\pi, 4\pi$, etc. radians; A_1P_1 is zero when $(\sin \omega t + \phi_1) = 0$, i.e. when $\omega t = -\phi_1, \pi - \phi_1, 2\pi - \phi_1, 3\pi - \phi_1$, etc. radians; and A_3P_3 is zero when $(\sin \omega t - \phi_2) = 0$, i.e. when $\omega t = \phi_2, \pi + \phi_2, 2\pi + \phi_2, 3\pi + \phi_2$, etc. radians, all of which are shown in fig. 5.16.

Since the OP_1 sine wave reaches any particular phase before the OP_2 sine wave, it is said to have a phase lead of ϕ_1 radians compared to the OP_2 sine wave. Similarly, the OP_3 sine wave has a phase lag of ϕ_2 radians compared to the OP_2 sine wave. These phase differences may be expressed as angles as above or, less often, in time units by dividing the angles by the angular velocity ω.

The 'arms' producing these sine waves are known as phasors, since an examination of them when stationary yields the phase relationship of the waves they produce. The length of the phasor also yields the amplitude in the diagrams shown so far.

Example 5.6 A voltage of sinusoidal waveform, amplitude 100 V, leads a current of sinusoidal waveform, amplitude 5 A, by $\pi/6$ radians. Calculate the instantaneous value of the voltage when the instantaneous value of the current is 2.5 A.

The equation describing the current waveform is $i = 5 \sin \omega t$, and the equation describing the voltage waveform is $v = 100 \sin (\omega t + \frac{\pi}{6})$

If the instantaneous value of current is 2.5 A, then $2.5 = 5 \sin \omega t$ and $\omega t = \arcsin 0.5$ (the angle having a sine equal to 0.5) $= \frac{\pi}{6}$ rad,

so that the instantaneous value of voltage is given by

$$v = 100 \sin \left(\frac{\pi}{6} + \frac{\pi}{6}\right) = 86.6 \text{ V}$$

Example 5.7 Calculate the frequency of a sinusoidally varying voltage of amplitude 10 V if its instantaneous value is 7.07 V at a time 2.5 ms after the beginning of a cycle.

The equation of the voltage waveform is $v = 10 \sin \omega t$ and so $7.07 = 10 \sin (\omega \times 2.5 \times 10^{-3})$

$$\text{Hence } \omega \times 2.5 \times 10^{-3} = \arcsin 0.707 = \frac{\pi}{4}$$

$$\text{Thus } \omega = \frac{\pi}{2.5 \times 4} \times 10^3$$

and since $\omega = 2\pi \times$ frequency,

$$\text{Frequency} = \frac{\pi}{2\pi \times 2.5 \times 4} \times 10^3 = 50 \text{ Hz}$$

Example 5.8 Determine the phase angle between two sinusoidally varying currents each of amplitude 1 A if the instantaneous values at a certain particular point in time are 0 and -1 A respectively. Clearly, when one current is passing through zero the other is at a

Addition of alternating voltages and currents using phasor diagrams

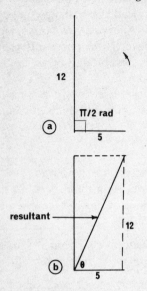

Figure 5.17

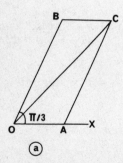

Figure 5.18a

negative peak and will pass through zero $\pi/2$ rad later. The phase angle is thus $\pi/2$ radians.

Instead of drawing waveforms each time we require to determine the resultant of alternating voltages or currents we need only draw the phasors which when rotating would produce the waveforms. We may then use standard geometrical and trigonometrical methods to obtain the resultant. The following examples should be studied carefully.

Example 5.9   Find the resultant of two voltages of peak value 12 V and 5 V respectively, when the 12 V peak voltage *leads* the other (a) by $\pi/2$ radians; (b) by $\pi/3$ radians.

(a) The phasor diagram is shown in fig. 5.17 (a). The 5 V phasor is drawn along the horizontal, the 12 V phasor is drawn at the angle $\pi/2$ radians (90°) in front of the 5 V phasor (the phasors rotate anticlockwise so 'in front' means at an angle of $\pi/2$ in an anticlockwise direction).

To determine the resultant, complete the rectangle as shown in fig. 5.17(b). The diagonal is the phasor which would produce the resultant, its length telling us the peak value of the resultant, its position telling us the phase relationship with the two component waveforms. The phasor diagram may be drawn to scale or we may use mathematical methods as follows:

By Pythagoras' Theorem, peak value of resultant = $\sqrt{(5^2 + 12^2)} = 13$

$$\tan \theta = \frac{12}{5}, \text{ thus } \theta = 67° \, 23' \text{ from tables (1.18 rad).}$$

The resultant is thus of peak value 13 V leading the waveform of peak value 5 V by 67° 23'.

NOTE that if the equation relating the instantaneous value v_1 of the 5 V peak voltage to its peak value is $v_1 = 5 \sin \omega t$, then the instantaneous value of the 10 V peak voltage v_2 is given by $v_2 = 10 \sin (\omega t + \pi/2)$, and since the resultant leads the 5 V peak voltage by 67° 23', i.e. 1.18 radians, the instantaneous value of the resultant, v_3, is given by

$$v_3 = 13 \sin (\omega t + 1.18)$$

(b) The phasor diagram is shown in fig. 5.18(a), where OA represents the 5 V phasor; OB represents the 12 V phasor; the angle between OB and OA is $\pi/3$ rad (60°); AC is drawn parallel and equal to OB; BC is drawn parallel and equal to OA.

The parallelogram OACB is completed and OC represents the peak value of the resultant voltage; COA is the angle by which the resultant *leads* the 5 V peak voltage represented by OA.

The magnitude of the resultant peak value, OC, may be determined *either* by drawing fig. 5.18(a) to scale (i.e. making the actual length of each line OA, OB proportional to the peak voltages they represent, for example, 5 cm and 12 cm respectively) and measuring the length of OC, *or* by using standard trigonometrical methods. Similarly, the phase angle COA between the resultant OC and the 5 V peak voltage may be measured directly on a scale diagram or determined by calculation.

The reader is left to draw a scale diagram; the calculation is as follows:

If OA is continued to some point X, then angle CAX = BOA (parallel sides BO and CA) so that

$$\text{angle OAC} = \pi - \text{CAX} = \pi - \text{BOA}$$
$$= \pi - \frac{\pi}{3} \text{ (since BOA} = \pi/3 \text{ rad)} = \frac{2\pi}{3} \text{ rad } (120°)$$

Using the cosine rule:
$$OC^2 = OA^2 + AC^2 - 2OA\,AC \cos OAC$$
$$= 25 + 144 - 2 \times 5 \times 12 \times \cos 2\pi/3$$
$$= 169 + 120 \cos \pi/3 \text{ (since } \cos 2\pi/3 = -\cos \pi/3)$$
$$= 169 + 120 \times 0.5$$
$$OC^2 = 229, \text{ so OC} = 15.13$$

The peak value of the resultant voltage is thus 15.13 V.

Angle COA may be calculated using the sine rule:

$$\frac{AC}{\sin COA} = \frac{OC}{\sin OAC}$$

$$\text{or } \sin COA = \frac{AC}{OC} \sin OAC = \frac{12}{15.13} \sin \frac{2\pi}{3}$$

$$= \frac{12 \times 0.866}{15.13} \text{ (since } \sin\frac{2\pi}{3} = \sin\frac{\pi}{3} = 0.866\text{)} = 0.6868$$

and COA = 0.757 rad (43° 23' from tables)

The resultant is thus a voltage of peak value 15.13 V leading the 5 V peak voltage by 0.757 rad.

Resultants (or component voltages making up a resultant) may also be found by *resolution of phasors*.

In fig. 5.17(b) we see that the 5 V peak voltage and 12 V peak voltage (leading the 5 V peak voltage by $\pi/2$ rad) have a resultant of peak value 13 V leading the 5 V peak voltage by 1.18 rad or 67° 23'.

Reversing the process, we can say that a voltage of peak value 13 V may be resolved into two components, one of peak value 5 V *lagging* the 13 V peak voltage by 67° 23' and one of peak value 12 V leading the 13 V peak voltage by 22° 37' (this angle being obtained by subtracting 67° 23' from 90°), i.e. the 13 V peak voltage as shown in fig. 5.17(b) has a *horizontal* component of 13 cos 67° 23' and a *vertical* component of 13 sin 67° 23' (or 13 cos 22° 37').

This technique will now be used in part (b) of the example in which the 12 V peak voltage leads the 5 V peak voltage by $\pi/3$ rad (60°). The answer should, of course, be the same as that previously obtained by completing the parallelogram and using trigonometrical methods.

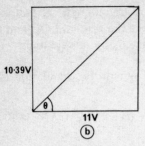

Figure 5.18b

In part (b) of the example the 5 V peak voltage is drawn along the horizontal and has *no* vertical component.

The 12 V peak voltage has a horizontal component of 12 cos $\pi/3$, i.e. 12 × 0.5, or 6 V and a vertical component of 12 sin $\pi/3$, i.e. 12 × 0.866, or 10.39 V.

The *total* horizontal component is thus 6 + 5 = 11 V, the *total* vertical component is 10.39 V. These components are shown in fig. 5.18 (b).

The resultant, by completion of the parallelogram and using Pythagoras' Theorem, is $\sqrt{(10.39^2 + 11^2)}$, i.e. $\sqrt{229}$ or 15.13 V, as before.

The phase angle θ by which the resultant leads the horizontal components (and therefore the 5 V peak voltage, which has *only* a horizontal component), is determined as follows:

$$\tan \theta = \frac{10.39}{11} = 0.9445, \text{ and } \theta = 43°\,23' \text{ as before.}$$

Summary

A unidirectional or direct current (d.c.) is one having a direction of flow which does not change; an alternating current (a.c.) has a direction of flow which changes, usually at regular intervals. A graph of current or voltage plotted against time is called the waveform of the current or voltage. There is a variety of waveforms including square, triangular and, the most common, the sinusoidal or sine wave, sometimes referred to as a sinusoid.

In a regular waveform (in which the pattern of change is repeated at regular intervals) the waveform which is regularly repeated is called one cycle of the total waveform. The time taken for one cycle is called the periodic time, or period, symbol T, and the number of cycles per second is called the frequency of the waveform, symbol f. Frequency is measured in cycles per second or hertz, symbol Hz, where one hertz is one cycle per second.

An alternating current or voltage has a number of possible values which are used to compare it with other alternating quantities. These include:
(1) the instantaneous value, which is the value at any instant;
(2) the maximum or peak value, which is the highest value reached in one direction;
(3) the peak-to-peak value or amplitude, which is equal to the sum of the maximum value in one direction and the maximum value in the opposite direction.
(4) the average or mean value, which is the rate of the constant magnitude direct current or voltage which would produce the same effect when that effect depends directly on the current or voltage (not the square of the current or voltage); and
(5) the root-mean-square, or r.m.s., value, which is the value of the constant magnitude direct current or voltage which would produce the same effect when that effect depends directly on the square of the current or voltage.

Alternating voltages and currents 97

The form factor of an alternating current or voltage is the r.m.s. value divided by the average value. For a sine wave the average value is equal to 0.637 × the maximum value, the r.m.s. value is equal to 0.707 × the maximum value and the form factor is equal to 1.11.

When two or more alternating currents or two or more alternating voltages act together the resultant current or voltage may be determined by adding together instantaneous values of the components to produce the waveform of the resultant and various values derived from this waveform. An easier method is to use phasors, where a phasor is a line drawn in such a way so as to indicate both the magnitude of a quantity and how the quantity is varying relative to other quantities (the phase relationship).

A phasor diagram consists of a number of such phasors and standard geometrical and trigonometrical methods may be used on the diagrams to derive resultant magnitudes and phase angles.

FURTHER WORKED EXAMPLES ON CHAPTER FIVE

5.10 Write down the relationship between frequency, f hertz, and periodic time, T seconds, of a regular alternating waveform.

$$\text{The relationship is } f = \frac{1}{T}$$

5.11 Calculate the frequency of a sinusoidal waveform if half a cycle occurs in 4 ms.

The periodic time is 2 × 4 ms, i.e. 8 ms.

$$\text{The frequency is thus } \frac{1}{8 \times 10^{-3}} \text{ (being 1/periodic time),}$$

$$\text{which equals } \frac{1000}{8}, \text{ i.e. 125 Hz.}$$

5.12 Calculate the time taken for 5 complete cycles of a sinusoidal alternating current of frequency 1 kHz.

$$\text{The periodic time is } \frac{1}{1000} \text{ s, i.e. 1 ms (being 1/frequency),}$$

and thus 5 cycles requires 5 × 1 ms, i.e. 5 ms.

5.13 State the relationship between the amplitude and peak-to-peak value of a regular alternating waveform.

The peak-to-peak value is equal to twice the amplitude of a regular alternating waveform.

5.14 Calculate the peak-to-peak value of a sinusoidal voltage of r.m.s. value 10 V.

The r.m.s. value of a sinusoidal voltage = 0.707 × peak value

so that peak value = $\dfrac{1}{0.707}$ × r.m.s. value

= 1.414 × r.m.s. value

and the peak-to-peak value = 2 × peak value
= 2.828 × r.m.s. value

So that in this case, peak-to-peak value = 2.828 × 10 = 28.28 V

5.15 The form factor of a sine wave is:

A. 1.414; B. 0.707; C. 0.637; D. 1.11.

All these figures are associated with a sine wave. A is the ratio peak value/r.m.s. value, B is the ratio r.m.s. value/peak value, C is the ratio average value/peak value and D is the form factor. D is therefore the right answer.

It is advisable to remember and *understand* definitions, rather than commit to memory a series of figures. In this way, the fact that form factor is equal to r.m.s. value divided by average value would come to mind and because the nature of r.m.s. and average values was understood and their relative values known, the likelihood of choosing the correct answer would be increased.

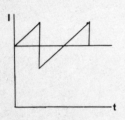

Figure 5.19

5.16 Fig. 5.19 shows the graph of values of a currrent plotted against time. The current is:

A. Direct and constant; B. Direct and varying; C. Alternating and constant; D. Alternating and varying.

A is obviously incorrect since the value of current is changing; C is impossible by definition (an alternating quantity cannot be constant); D is incorrect since, again by definition, alternating means 'changing direction' – this current is not changing direction. B is the correct answer.

5.17 An alternating voltage of peak-to-peak value 20 V is acting in phase with an alternating voltage of r.m.s. value 10 V. If both voltages are sinusoidal the resultant voltage peak value is:

A. 30 V; B. 17.07 V; C. 24.14 V; D. 34.14 V.

To obtain A, the peak-to-peak value of 20 V has been added to an r.m.s. value of 10 V. If dissimilar values of in-phase voltages are added the answer is meaningless. To obtain B, the r.m.s. value of the first voltage (which is 0.707 × peak, i.e. 0.707 × 20/2) has been added to the r.m.s. value of the second voltage; this answer is therefore the r.m.s. value of the resultant. Answer D results from adding a peak-to-peak value (20 V) to a peak value (1.414 × 10 V) and is incorrect. The correct answer C is obtained by adding the peak value of the first voltage (20/2 V) to the peak value of the second (1.414 × 10 V) to give 10 + 14.14, i.e. 24.14 V.

Alternating voltages and currents 99

5.18 The resultant voltage of the three voltages shown in the phasor diagram (fig. 5.20) is of value:

A. 27 V; B. 17 V; C. 19.21 V; D. 13 V.

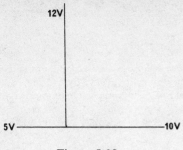

Figure 5.20

To obtain A, all three voltages have been added without taking into account that they are not in phase. To obtain answer B, 5 V has been subtracted from 10 V (which is correct since they are in opposition) but the resultant 5 V has then been added to 12 V; this is incorrect since they are not in phase. Answer C has been obtained by adding the 10 V to the 5 V to give 15 V, which is incorrect since they are not in phase, and then correctly summing the 15 V and the 12 V from the phasor diagram (fig. 5.21).

Answer D is correct. The horizontal component of the resultant is 10 V − 5 V, since they are in opposition, the vertical component is 12 V and the resultant, from fig. 5.22, is thus 13 V.

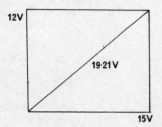

Figure 5.21

5.19 The instantaneous value of a sinusoidally varying current is given by $i = 20(\sin \omega t + \pi/6)$.
Select the *incorrect* statement from the following:

A. The peak value is 20 A; B. The phase angle is $\pi/6$ rad leading; C. The r.m.s. value is 14.14 A; D. The phase angle is $\pi/6$ rad lagging.

The peak value of this current is 20 A, its r.m.s. value (0.707×20) is 14.14 A. Therefore A and C are correct. The phase angle is $\pi/6$ rad and the plus sign preceding it in the equation tells us the current is *leading*. Answer D is thus incorrect and answer C correct. Since we are asked to select the *incorrect* statement, D is the right answer to the question.

This type of question emphasises the need for careful reading of problems.

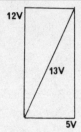

Figure 5.22

SELF-ASSESSMENT EXERCISE 5

Possible marks

1. Express the periodic time T seconds of a sine wave in terms of its frequency f hertz. (3)

2. Calculate the frequency of a sine wave if 4.5 cycles occur in 10 ms. (3)

3. Determine the periodic time of a regular alternating voltage of frequency 12 kHz. (3)

4. Determine the peak value, the r.m.s. value and the phase angle of an alternating current having an instantaneous value given by $i = 12 \sin(\omega t - \pi/3)$. (3)

5. A sinusoidally alternating voltage has an r.m.s. value of 15 V and a phase angle of 45° lagging. Write down an expression for its instantaneous value v volts in terms of time t seconds and angular velocity ω rad/s. (3)

6. The r.m.s. value of a sinusoidal voltage of peak value 100 V is: A. 141.4 V; B. 70.7 V; C. 63.7 V; D. 111 V. (3)

7. The peak-to-peak value of a square wave current of average value (over half a cycle) of 5 A is: A. 10 A; B. 7.07 A; C. 5 A; D. 14.14 A. (3)

8. The instantaneous values of two alternating currents, i_1, i_2 are given by:
$i_1 = 10 \sin(\omega t + \pi/4)$ and $i_2 = 15 \sin(\omega t - \pi/3)$.
The phase angle between these currents is:
A. $\pi/12$; B. $-\pi/12$; C. $-7\pi/12$; D. $7\pi/12$. (3)

9. An alternating current of r.m.s. value 4 A leads a second current, alternating in the same manner and at the same frequency, by $\pi/2$ rad. The resultant current has an r.m.s. value of 5 A. The r.m.s. value of the second current is:
A. 3 A; B. 1 A; C. 9 A; D. impossible to calculate without further information. (3)

10. A sinusoidal voltage has a peak value of 20 V and a phase angle of $\pi/6$ rad leading. If a phasor diagram is drawn in the normal way with the reference along the horizontal, the r.m.s. value of the horizontal component of this voltage is:
A. 10 V; B. 17.32 V; C. 7.07 V; D. 12.24 V. (3)

11. Calculate the frequency of a sinusoidally varying voltage of amplitude 15 V if its instantaneous value reaches 12 V for the first time in the cycle at 1.6 ms after the beginning of the cycle. (14)

12. A sinusoidal current of r.m.s. value 5 A leads a sinusoidal voltage of r.m.s. value 50 V by $\pi/5$ rad. Calculate the instantaneous value of the current when the voltage has an instantaneous value of 33 V. (14)

13. The instantaneous values of two voltages, v_1, v_2 are given by:
$v_1 = 20 \sin(\omega t + \pi/5)$ and $v_2 = 10 \sin(\omega t - \pi/3)$.
Calculate the resultant of these voltages using a phasor diagram and applying trigonometrical methods. (14)

14. Calculate the resultant of the voltages of question 13 using phasor resolution. (14)

15. Two alternating currents combine in a circuit to give a resultant current having an instantaneous value given by:
$i = 15 \sin(\omega t + \pi/3)$
The r.m.s. value of one current is 10 A, its phase angle $\pi/2$ leading. Calculate the peak value and the phase angle of the other current. Use either trigonometrical methods or phasor resolution. (14)

Answers

SELF-ASSESSMENT EXERCISE 5

Marks

1. $T = \dfrac{1}{f}$ (3)

2. 4.5 cycles occur in 10 ms, so
1 cycle occurs in $\dfrac{10}{4.5}$ ms (the periodic time), i.e. 2.22 ms.

Frequency $= \dfrac{1}{2.22 \times 10^{-3}} = \dfrac{1000}{2.22} = 450$ Hz (3)

3. Periodic time $= \dfrac{1}{12 \times 10^3} = \dfrac{1}{12}$ ms $= \dfrac{1000}{12} \mu s = 83.33 \mu s$ (3)

4. From the equation $i = 12 \sin(\omega t - \dfrac{\pi}{3})$

by examination, peak value = 12 A (1)

therefore r.m.s. value = $12 \times 0.707 = 8.48$ A (1)

by examination, phase angle = $\dfrac{\pi}{3}$ lagging (1)

5. $45° = \frac{\pi}{4}$ rad; peak value $= 1.414 \times 15$ V (2)

The expression is therefore $v = 1.414 \times 15 \sin(\omega t - \frac{\pi}{4})$

$= 21.21 \sin(\omega t - \frac{\pi}{4})$ (1)

6. R.M.S. value is 0.707×100 V, i.e. 70.7 V

B is the correct answer (3)

7. The average value of a square wave over half a cycle is the same as the peak value.

Hence peak value $= 5$ A, and peak-to-peak value $= 2 \times 5 = 10$ A

A is the correct answer (3)

8. 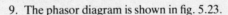 Current i_1 leads the reference by $\frac{\pi}{4}$ rad

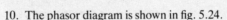 Current i_2 lags the reference by $\frac{\pi}{3}$ rad

The angle between the phasors representing these currents is therefore $(\frac{\pi}{4} + \frac{\pi}{3})$ rad, i.e. $\frac{7\pi}{12}$

Since we are not asked which current leads the other the sign is not relevant.

D is the correct answer. (3)

Figure 5.23

9. The phasor diagram is shown in fig. 5.23.

The r.m.s. value of the unknown current is given by

$\sqrt{(5^2 - 4^2)}$ by Pythagoras, i.e. 3 A

A is the correct answer. (3)

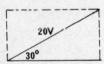

Figure 5.24

10. The phasor diagram is shown in fig. 5.24.

The horizontal component is therefore $20 \cos 30°$ V, i.e. 17.32 V (peak value), or $20 \times 0.707 \cos 30°$ V, i.e. 12.24 V (r.m.s. value).

The r.m.s. value is thus 12.24 V and D is the correct answer (3)

11. The instantaneous value is given by $V = 15 \sin \omega t$, where $\omega = 2\pi \times$ frequency and t is time (seconds). (4)

When $t = 1.6$ ms, $V = 12$ V, and inserting these values in the equation for the instantaneous value we have

$12 = 15 \sin(\omega \times 1.6 \times 10^{-3})$, or $\sin(\omega \times 1.6 \times 10^{-3}) = \frac{12}{15} = 0.8$

From tables the angle having a sine equal to 0.8 is 0.9273 rad (53° 8').

Thus $\omega \times 1.6 \times 10^{-3} = 0.9273$, and since $\omega = 2\pi f$ where f is frequency (Hz), (6)

$2\pi f \times 1.6 \times 10^{-3} = 0.9273$ and $f = \frac{0.9273 \times 10^3}{2\pi \times 1.6} = 92.24$ Hz (4)

12. The instantaneous value of voltage is given by $V = 1.414 \times 50 \sin \omega t$ (3)

and of the current by $i = 1.414 \times 5 \sin(\omega t + \frac{\pi}{5})$ (3)

When the instantaneous value of voltage is 33 V,

$33 = 1.414 \times 50 \sin \omega t$ and $\sin \omega t = \frac{33}{1.414 \times 50} = 0.4668$ (4)

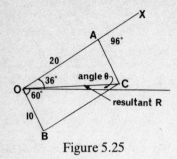

Figure 5.25

so that $\omega t = 27° 50'$ (0.4858 rad),

and the instantaneous value of current

$$= 1.414 \times 5 \sin (27° 50' + 36°) \quad (\tfrac{\pi}{5} \text{ rad is } 36°)$$
$$= 1.414 \times 5 \sin 63° 50' = 1.414 \times 5 \times 0.8976 = 6.346 \text{ A} \qquad (4)$$

13. The phasor diagram is shown in fig. 5.25, where R is the resultant and θ the phase angle of the resultant. The value of R may be found from the triangle OAC in fig. 5.25:

$$\text{A}\hat{\text{O}}\text{B} = 36° + 60° = 96°, \text{ so that } \text{X}\hat{\text{A}}\text{C} = 96° \text{ (AC parallel to OB)}.$$
$$\text{O}\hat{\text{A}}\text{C} = 180° - 96° = 84° \qquad (3)$$
$$\text{and } R^2 = 20^2 + 10^2 - (2 \times 20 \times 10 \cos 84°)$$
$$= 500 - (400 \times 0.1045) = 458.2$$
$$\text{so that } R = 21.4 \qquad (4)$$

To find θ we may use the sine rule as follows:

$$\frac{\text{AC}}{\sin \text{A}\hat{\text{O}}\text{C}} = \frac{R}{\sin \text{O}\hat{\text{A}}\text{C}},$$

i.e. $\sin \text{A}\hat{\text{O}}\text{C} = \dfrac{\text{AC} \sin \text{O}\hat{\text{A}}\text{C}}{R} = \dfrac{10 \sin 84°}{21.4} = 0.4647 \qquad (4)$

so that $\text{A}\hat{\text{O}}\text{C} = 27° 42'$, and since $\text{A}\hat{\text{O}}\text{C} = 36° - \theta$,

$$\theta = 36° - \text{A}\hat{\text{O}}\text{C} = 36° - 27° 42' = 8° 18' \qquad (3)$$

The resultant is of peak value 21.4 V and leads by 8° 18' (0.1449 rad).

The instantaneous value is given by $V = 21.4 \sin (\omega t + 0.1449)$

NOTE that had the phasor diagram been drawn showing a lagging phase angle the mathematics would have yielded a negative value for θ, indicating that it was the opposite to the sense drawn, i.e. below the horizontal reference rather than above it.

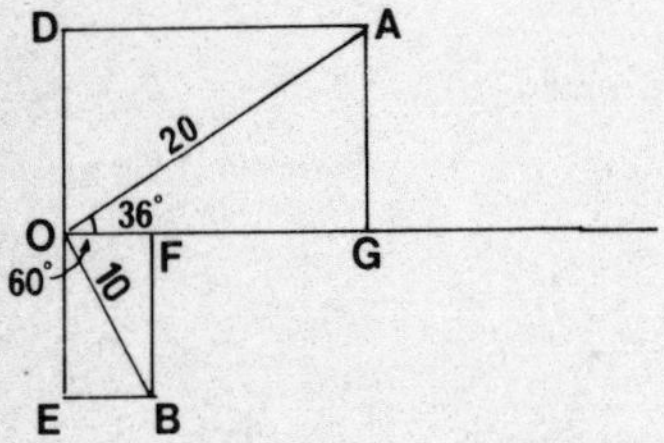

14. The phasors are drawn and resolved into vertical and horizontal components as shown in fig. 5.26.

Voltage represented by OA may be resolved into a horizontal component OG and a vertical component OD as follows:

$$\text{OG} = 20 \cos 36° = 16.18; \text{OD} = 20 \sin 36° = 11.76 \qquad (1)(1)$$

Similarly, the voltage represented by OB may be resolved into OF horizontally and OE vertically:

$$\text{OF} = 10 \cos 60° = 5; \text{OE} = 10 \sin 60° = 8.66 \text{ (acting negatively relative to OD)} \qquad (1)(1)$$

Total horizontal component $= \text{OG} + \text{OF} = 16.18 + 5 = 21.18 \qquad (2)$

Total vertical component $= \text{OD} - \text{OE} = 11.76 - 8.66 = 3.1$ (positive) (see fig. 5.27) $\qquad (2)$

and the resultant R is given by $\sqrt{(3.1^2 + 21.18^2)} = 21.4$, as before $\qquad (2)$

θ is found by $\tan \theta = \dfrac{3.1}{21.18}$; hence $\theta = 8° 18'$, as before. $\qquad (4)$

15. The resultant current is of peak value 15 A and has a leading phase angle of 60°. Its phasor may be drawn as shown in fig. 5.28, and since one of the component currents has a peak value of 1.414×10, i.e. 14.14 A and leads by 90°, its phasor may be added to the diagram, the parallelogram completed and the phasor of the second component current shown as in fig. 5.29.

Figure 5.26

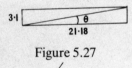

Figure 5.27

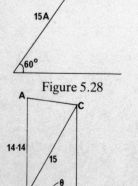

Figure 5.28

Figure 5.29

OB is the phasor of the second component current; θ is the phase angle of the second component current.

Phasor resolution is the neater method of solution as follows: Vertical component of OC = vertical component of OA + vertical component of OB.

$$15 \sin 60° = 14.14 - OB \sin θ \tag{4}$$

NOTE that since OB has been assumed to lie below the horizontal, its vertical component is negative with respect to the vertical component of OA.

$$OB \sin θ = 14.14 - 15 \sin 60° = 1.15$$

Similarly, horizontal component of OC = horizontal component of OB (since OA has no horizontal component),

$$\text{so that } 15 \cos 60° = OB \cos θ, \text{ and } OB \cos θ = 7.5 \tag{4}$$

We now have OB sin θ = 1.15 and OB cos θ = 7.5

$$\text{Hence, by division } \frac{\sin θ}{\cos θ} = \frac{1.15}{7.5},$$

$$\text{i.e. } \tan θ = 0.1533 \text{ and } θ = 8° 43'. \tag{3}$$

Since OB cos θ = 7.5, i.e. OB cos 8° 43' = 7.5,

$$OB = \frac{7.5}{\cos 8° 43'} = \frac{7.5}{0.9885} = 7.59 \tag{3}$$

The peak value of the second component current is 7.59 A and its phase angle 8° 43' (0.1521 rad) lagging the reference or 68° 43' lagging the resultant.

6 Single-phase a.c. circuits

Topic area: F

General objective The expected learning outcome is that the student understands the behaviour of simple series a.c. circuits.

Specific objectives The expected learning outcome is that the student:
8.1 States that in a purely resistive circuit I is in phase with V, and that in a purely inductive circuit I lags V by 90° or $\pi/2$ radians and that in a purely capacitive circuit I leads V by 90° or $\pi/2$.
8.2 Draws the phasor diagrams and relative voltage and current waveforms relating to 8.1.
8.3 Describes inductive reactance and capacitive reactance in terms of impeding the flow of an alternating current.
8.4 States inductive reactance as: $X_L = V_L/I_L = 2\pi fL = \omega L$.
8.5 States capacitive reactance as: $X_C = V_C/I_C = 1/2\pi fC = 1/\omega C$.
8.6 Applies the equation in 8.4 and 8.5 to simple problems.

Resistance, reactance and impedance

When a voltage is applied to a conductive circuit an electric current flows. However good the conductors in the circuit are there is always some opposition to current flow and, in d.c. circuits, dividing the steady state applied voltage by the resultant steady state current gives a measure of this opposition. The quantity is called resistance and is due to the molecular structure of the circuit materials and to their physical dimensions.

The words 'steady state' are used because, as was shown in chapters 2 and 4, there is an opposition to changing voltage in capacitive circuits and to changing current in inductive circuits. The 'steady state' is the period which commences once all the initial or transient changes are complete. In a.c. circuits there is still the transient period when the voltage is first applied but even in the steady state the voltages and currents in the circuit are changing all the time.

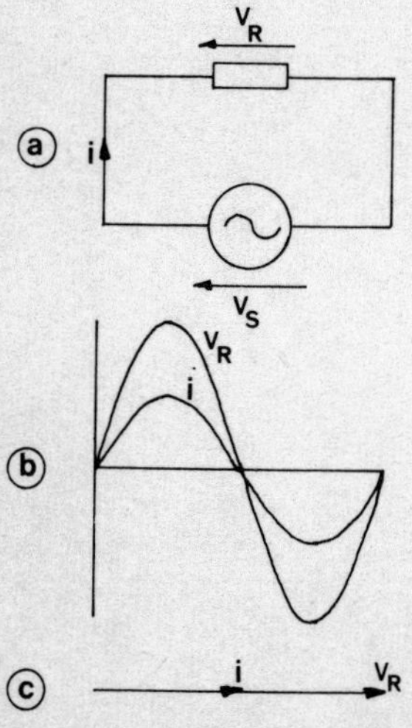

In circuits containing inductance and capacitance there will be a continual opposition due, not to physical dimensions or structure of conductors, but to the effects of electromagnetic induction in the one case and the fact that a capacitor cannot charge instantaneously in the other. Opposition to alternating current flow due to inductance or capacitance alone is called *reactance* (the circuit *reacts* to change), the special names being inductive reactance and capacitive reactance respectively. The combined effect of resistance and reactance (for the effects of resistance apply equally in a.c. or d.c. circuits) is called *impedance*.

Figure 6.1

Resistance in a.c. circuits Resistance is unaffected by changing voltages and currents (except in certain rather special cases at very high frequencies), so that the relationship

$$v = iR$$

where v, i and R represent voltage, current and resistance measured in volts, amperes and ohms respectively, remains the same. As the applied voltage changes, its instantaneous value varying with time, the current does the same, reaching its maximum value at the same instant that the voltage reaches its maximum value and passing through zero as the voltage passes through zero. Waveform diagrams and phasor diagrams are shown in fig. 6.1.

In a purely resistive circuit the current is said to be *in phase* with the applied voltage.

Inductive reactance in a.c. circuits

Changing voltage applied to an inductive coil sets up a changing current and, in turn, a changing magnetic flux linking the coil. A changing induced e.m.f. is thus established across the coil. The value of this e.m.f. is given by

$$e = -L\frac{\mathrm{d}i}{\mathrm{d}t}$$

where e is the induced e.m.f. (volts), L is the coefficient of self-inductance (henrys) and $\mathrm{d}i/\mathrm{d}t$ represents the rate of change of current with time (amperes/second).

It is not in fact possible to have a purely inductive coil since the windings must have resistance, but if we consider a coil in which the resistance is so small that it can be ignored (in comparison with the effects of the coil inductance), the applied voltage may be assumed equal (and opposite) to the induced e.m.f. Denoting this applied voltage by v_L we can write

$$v_L = L\frac{\mathrm{d}i}{\mathrm{d}t}$$

i.e. the coil voltage $= L \times$ rate of change of current with time.

The applied voltage does not change exactly as the current as it does in a purely resistive circuit, but as the *rate of change* of the current. This produces waveforms and a phasor diagram as shown in fig. 6.2.

Examine the waveform diagram shown in fig. 6.2b. At points A and C the current is at a peak value, negative at A, positive at C and is not changing at these instants. The voltage is zero at these points. At point B, however, the current, although having a zero value, is reversing its direction of flow and its *rate of change* is a *maximum* as is the applied voltage. Similarly, at point D the current is again reversing its direction, this time in an opposite sense, and its rate of change is again a maximum as is the voltage.

Denoting regions above the time axis as 'positive' and below as 'negative', the voltage has a positive peak value as the current reverses from negative to positive and a negative peak value as the current reverses from positive to negative.

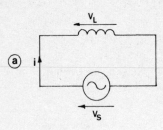
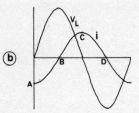
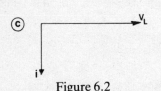

Figure 6.2

As can be seen, the opposition to changing current *delays* the current change, the current *lagging* behind the voltage. It is found that as the resistance of an inductive circuit is made smaller and smaller this delay or *lag* becomes larger and approaches the time taken for a quarter of the periodic time (the time taken for a full cycle). Thus for a pure inductance the lag would be a quarter of a cycle or, using angular measure, $\pi/2$ radians. The phasor diagram is shown in part (c) of fig. 6.2.

In the circuit shown, assuming the coil has zero resistance, the inductive reactance is given by dividing the supply voltage by the current. Inductive reactance has the symbol X_L and is measured in ohms (as is resistance).

Inductive reactance of a particular inductor depends upon the self-inductance of the coil as might be expected and on the frequency of the supply, the greater the frequency the greater the reactance. This, too, might be expected since it is the alternating nature of the voltage that causes the opposition.

It can in fact be shown that $X_L = 2\pi f L$, or $X_L = \omega L$

where ω is the angular velocity of the phasors, as explained in the previous chapter ($\omega = 2\pi f$).

Capacitive reactance in a.c. circuits

The process of charging a capacitor takes time so that a voltage across it cannot be established instantly. A capacitor thus delays a changing voltage in a manner similar to that in which an inductor delays a changing current. The equation connecting charge capacitance and voltage is

$$q = Cv$$

where q is the charge (coulombs), C the capacitance (farads) and v the voltage (volts). Lower-case symbols are used to indicate instantaneous values.

If $q = Cv$, then $\dfrac{dq}{dt} = C\dfrac{dv}{dt}$

i.e. rate of change of charge with time = $C \times$ rate of change of voltage with time.

Electric current is the flow of electric charge and is measured in charge per second. The rate of change of charge with time is thus electric current.

$$i = \frac{dq}{dt} = C\frac{dv}{dt}, \text{ where } i \text{ represents current (amperes)}.$$

Thus current = capacitance × rate of change of voltage.

(Compare the inductive case where

voltage = inductance × rate of change of current.)

Single-phase a.c. circuits 107

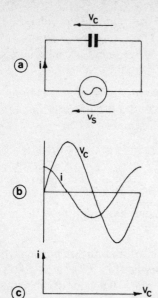

Figure 6.3

In a.c. circuits, voltage and current will not rise and fall simultaneously as with resistance but there will be a *phase shift* as there is with inductance.

The waveforms and phasor diagram are shown in fig. 6.3.

As can be seen, the voltage waveform reaches its maximum values a quarter of a cycle *after* the current waveform reaches its maximum values. Current is said to *lead* voltage, the lead being a quarter of a cycle, or, in angular measure $\pi/2$ rad.

A useful aid to memory is the word 'civil':

CIVIL

capacitance – current – voltage – current – inductance

which reminds us that for a capacitive circuit (C), current I leads voltage V, and voltage V leads current I in an inductive (L) circuit.

Capacitive reactance is in opposition to alternating current due to capacitance alone. In the circuit shown it is given by dividing the supply voltage by the current and it is measured in ohms. The symbol is X_C.

In a d.c. circuit current flows in the capacitor leads until the capacitor is fully charged. The current then falls to zero. At the beginning of the charging process the current is high and reduces as charge accumulates (the negative plate repelling further electrons). The time taken to fully charge the capacitor depends upon, among other things, the capacitance, the larger the capacitance the longer being the charging time (since capacitance is a measure of the charge that can be stored per volt).

In an a.c. circuit the current is rising and falling and at regular intervals its direction of flow is reversed. The capacitor is therefore in a constant state of being charged or discharged and current flows in the leads all the time. The size of the current at any time depends, of course, on the applied voltage but also on the capacitance and on the supply frequency.

The larger the capacitance the longer it takes to charge and the capacitor is further away from being fully charged, so that the current is in the 'beginning-to-charge' state, i.e. it is larger. Similarly, the higher the frequency, the more rapid is the rate of reversal and again the capacitor is further away from being fully charged, the current again being larger.

The opposition to current flow is thus reduced as the capacitance or the frequency is increased and capacitive reactance is *indirectly* proportional to capacitance and frequency.

It can be shown that $X_C = \dfrac{1}{2\pi f C}$ or $X_C = \dfrac{1}{\omega C}$, where $\omega = 2\pi f$.

Example 6.1 Calculate the current flowing in a 150 Ω resistor when a 240 V alternating voltage is applied across it.

Current = voltage/resistance = $\dfrac{240}{150}$ = 1.6 A

The current is 1.6 A.

Note that if it is not stated otherwise, alternating voltages and currents are given in r.m.s. values. The voltage here is thus 240 V r.m.s. so that the current too is the r.m.s. value. Care must be taken to calculate the *required* value as stated in the problem.

Example 6.2 Calculate the inductive reactance of a 10 H inductor at 50 Hz and the capacitive reactance of a 0.2 μF capacitor at the same frequency.

$$\text{For the inductor, } X_L = 2\pi f L = 2\pi \times 50 \times 10 = 3141.6 \text{ }\Omega$$

$$\text{For the capacitor, } X_C = \frac{1}{2\pi f C} = \frac{10^6}{2\pi \times 50 \times 0.2} = 15\,915 \text{ }\Omega$$

The inductive reactance is 3141.6 Ω and the capacitive reactance is 15 915 Ω.

Example 6.3 Calculate the peak value of the current flowing when a 150 V, 50 Hz sinusoidal voltage is applied to a 0.2 H inductor. The resistance of the inductor may be ignored. What value of capacitance would give the same value of current if this voltage were applied to it?

$$\text{Inductive reactance} = 2\pi f L = 2\pi \times 50 \times 0.2 = 62.83 \text{ }\Omega$$

$$\text{Current} = \text{voltage/inductive reactance} = \frac{150}{62.83} = 2.39 \text{ A}$$

This is the r.m.s. value of current.

Peak value of current = 1.414 × r.m.s. value of current
$$= 1.414 \times 2.39 = 3.38 \text{ A}$$

The peak value of current is 3.38 A.

For the same value of current at the same value of voltage the capacitive reactance must equal the inductive reactance, 62.83 Ω.

$$\text{Capacitive reactance} = \frac{1}{2\pi f C}, \text{ i.e. } 62.83 = \frac{1}{2\pi \times 50 \times C}$$

$$C = \frac{1}{2\pi \times 50 \times 62.83} = 50.66 \times 10^{-6} = 50.66 \text{ }\mu\text{F}$$

The value of capacitance is 50.66 μF

Example 6.4 Calculate the frequency at which the reactance of a 5 H inductor is equal to the reactance of a 0.2 μF capacitor.

$$\text{Inductive reactance} = 2\pi f L; \text{ capacitive reactance} = \frac{1}{2\pi f C}$$

$$\text{When these are equal, } 2\pi f L = \frac{1}{2\pi f C} \text{ and } f^2 = \frac{1}{4\pi^2 LC}$$

Inserting the given values:

$$f^2 = \frac{10^6}{4\pi^2 \times 5 \times 0.2} = 2.533 \times 10^4 \text{ and } f = 159.15 \text{ Hz}$$

The frequency is 159.15 Hz.

Example 6.5 When a sinusoidal voltage of peak value 400 V at a frequency 75 Hz is applied to a certain resistor, the current flowing is 125 mA r.m.s. An inductor and then a capacitor is placed in turn across the same supply and the same current flows.

Calculate the inductance of the inductor, assuming its resistance is negligible, and the capacitance of the capacitor.

$$\text{r.m.s. value of voltage} = \frac{400}{1.414} = 282.88$$

$$\text{r.m.s. value of current} = 0.125 \text{ A}$$

$$\text{hence resistance} = \frac{282.88}{0.125}$$

$$= 2263.1 \, \Omega$$

For the same current at the same voltage the inductive reactance and this value of resistance must be equal, i.e.

$$2263.1 = 2\pi \times 75 \times L \text{ and } L = \frac{2263.1}{2\pi \times 75} = 4.8 \text{ H}$$

The capacitive reactance also equals 2263.1 Ω.

$$\frac{1}{2\pi \times 75 \times C} = 2263.1$$

$$\text{and } C = \frac{1}{2263.1 \times 2\pi \times 75} \text{ F} = 0.94 \, \mu\text{F}$$

The inductance of the inductor is 4.8 H and the capacitance of the capacitor is 0.94 μF.

Specific objectives

The expected learning outcome is that the student:
8.7 Draws phasor diagrams corresponding to L-R and C-R series circuits.
8.8 Determines triangles derived from the phasor diagrams of 8.7.
8.9 Defines impedance as $Z = V/I$.
8.10 Derives impedance triangles from voltage triangles.
8.11 Shows that $Z^2 = R^2 + X^2$ and that $\tan \phi = X/Z$ and $\cos \phi = R/Z$.
8.12 Applies equations in 8.9 and 8.11 to the solution of single-branch L-R series circuits at power and radio frequencies.

Series *L-R* and *C-R* circuits

In a purely resistive circuit there is no phase shift between voltage and current, the two are in phase. In a purely reactive circuit (inductive or capacitive) there is a quarter of a cycle phase shift, the current either leading or lagging the voltage depending upon whether the circuit is capacitive or inductive. When resistance and reactance are combined the phase shift will lie *between* zero and a quarter of a cycle depending upon which component has the greater effect.

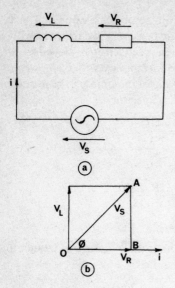

Figure 6.4

First consider a series L-R circuit as shown in figure 6.4.

The supply voltage, V_S, has two components, V_L across the inductance and V_R across the resistance. V_R is in phase with the current and V_L leads the current by $\pi/2$ rad as shown. V_L and V_R are summed by completing the rectangle, the diagonal representing the supply voltage phasor. The phase angle ϕ lies between zero and $\pi/2$ rad as shown. If the inductive reactance is much larger than the resistance, V_L will be much larger than V_R, since the same current flows in both components and the phase angle will be closer to $\pi/2$ rad. Conversely, if the resistance is much larger than the reactance, the phase angle will be closer to zero.

As was stated earlier, an inductor always has some resistance so that if the circuit represents an inductor in series with a resistor, the inductor resistance will be added to the resistance of the resistor when calculations are carried out. Measurement in a practical circuit using instruments will not give V_L but a combination of V_L and whatever part of V_R is required for the inductor resistance. Practically, of course, the resistance and inductance of the inductor could not be separated.

Fig. 6.5 shows the waveform and phasor diagrams of a simple series C-R circuit.

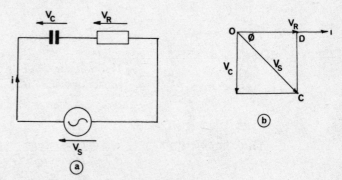

Figure 6.5

Here the supply voltage is the phasor sum of the component voltages V_R and V_C as shown, V_R being in phase with the current, V_C lagging the current by $\pi/2$ rad. Again, the phase angle between supply voltage and current will be determined by the respective values of V_C and V_R, which, in turn, are determined by the values of capacitance and resistance in the circuit. If X_C is much larger than R, the phase angle will approach $\pi/2$ rad; if R is much larger than X_C, the phase angle will be closer to zero.

Impedance

Impedance is opposition to alternating current flow due to the combined effect of resistance and reactance. Since resistance and reactance have different effects on voltage and current relationships, it is not a simple matter of directly adding the two quantities to obtain impedance. They must be summed using phasor methods as shown in fig. 6.6.

Part of the phasor diagrams in figs 6.4 and 6.5 are redrawn in fig. 6.6. Fig. 6.6a shows the voltage relationships in a series L-R circuit, fig. 6.6b shows these relationships in a series C-R circuit. The triangle showing the relationships in fig. 6.6a is triangle OAB in fig. 6.4b, and the triangle in fig. 6.6b is triangle OCD in fig. 6.5b.

Applying Pythagoras' Theorem to the voltage triangle in fig. 6.6a,

$$V_S^2 = V_L^2 + V_R^2$$

and to the voltage triangle in fig. 6.6b,

$$V_S^2 = V_C^2 + V_R^2$$

so that for a series L-R circuit $V_S = \sqrt{(V_L^2 + V_R^2)}$

and for a series C-R circuit $V_S = \sqrt{(V_C^2 + V_R^2)}$

Denoting current by i, the voltages V_S, V_L, V_R and V_C can be written:

$V_S = iZ$, where Z is the total circuit impedance;
$V_L = iX_L$; $V_R = iR$; and $V_C = iX_C$

so that for a series L-R circuit:

$$V_S^2 = V_L^2 + V_R^2 \text{ and } (iZ)^2 = (iX_L)^2 + (iR)^2,$$

which gives $Z = \sqrt{(X_L^2 + R^2)}$

i.e. the circuit impedance is the square root of the sum of the square of the inductive reactance and the square of the resistance.

For a series C-R circuit $V_S^2 = V_C^2 + V_R^2$ and $(iZ)^2 = (iX_C)^2 + (iR)^2$ so that $Z = \sqrt{(X_C^2 + R^2)}$

and the impedance of a series C-R circuit is the square root of the sum of the square of the capacitive reactance and the square of the resistance.

The voltage triangles in figs 6.6a and 6.6b can be redrawn as impedance triangles as shown directly beneath the voltage triangles in the figure. The voltage and impedance triangles for each circuit to which they refer are similar; to obtain the impedance triangle each side of the voltage triangle is divided by the current.

The difference between the triangles is, of course, that the voltage triangle represents phasor quantities (i.e. quantities having a phase or time relationship), the impedance triangle does not, since impedance and reactance are not phasor quantities.

From the impedance triangles an equation giving the value of the phase angle ϕ can be obtained:

For the series L-R circuit (fig. 6.6a). $\sin\phi = \dfrac{V_L}{V_S}$ or $\dfrac{X_L}{Z}$,

$\cos\phi = \dfrac{V_R}{V_S}$ or $\dfrac{R}{Z}$ and $\tan\phi = \dfrac{V_L}{V_R}$ or $\dfrac{X_L}{R}$

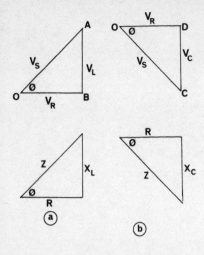

Figure 6.6

For the series C-R circuit (fig. 6.6b), $\sin\phi = \dfrac{V_C}{V_S}$ or $\dfrac{X_C}{Z}$,

$$\cos\phi = \dfrac{V_R}{V_S} \text{ or } \dfrac{R}{Z} \text{ and } \tan\phi = \dfrac{V_C}{V_R} \text{ or } \dfrac{X_C}{R}$$

A summary of these relationships follows:

Series L-R circuit

$V_S = \sqrt{(V_L^2 + V_R^2)}$
$Z = \sqrt{(X_L^2 + R^2)}$
$\sin\phi = \dfrac{V_L}{V_S}$ or $\dfrac{X_L}{Z}$
$\cos\phi = \dfrac{V_R}{V_S}$ or $\dfrac{R}{Z}$
$\tan\phi = \dfrac{V_L}{V_R}$ or $\dfrac{X_L}{R}$

Series C-R circuit

$V_S = \sqrt{(V_C^2 + V_R^2)}$
$Z = \sqrt{(X_C^2 + R^2)}$
$\sin\phi = \dfrac{V_C}{V_S}$ or $\dfrac{X_C}{Z}$
$\cos\phi = \dfrac{V_R}{V_S}$ or $\dfrac{R}{Z}$
$\tan\phi = \dfrac{V_C}{V_R}$ or $\dfrac{X_C}{R}$

The following worked examples should be studied carefully.

Example 6.6 In a series C-R circuit the supply voltage is 240 V, 50 Hz. The voltage across the resistor is 115 V. Calculate the voltage across the capacitor.

$$\text{Using } V_S^2 = V_C^2 + V_R^2, (240)^2 = V_C^2 + (115)^2$$
$$\text{and } V_C^2 = (240)^2 - (115)^2, \text{ so that } V_C = 210.65$$

The voltage across the capacitor is 210.65 V

Example 6.7 An inductor of inductance 10 H and resistance 200 Ω is connected in series with a 500 Ω resistor, the series circuit then being connected across a 115 V, 50 Hz sinusoidal supply. Calculate the:
   (a) inductive reactance of the inductor;
   (b) circuit impedance;
   (c) circuit current;
   (d) voltage across the resistor.

(a) Inductive reactance $= 2\pi fL = 2\pi \times 50 \times 10 = 3141.6$ Ω

(b) Circuit impedance $= \sqrt{(X_L^2 + R^2)} = \sqrt{(3141.6^2 + 700^2)} = 3218.6$ Ω

(Note: the *total* circuit resistance is 500 + 200, i.e. 700 Ω)

(c) Circuit current = supply voltage/circuit impedance

$$= \dfrac{115}{3218.6} = 0.036 \text{ A}$$

(d) Voltage across resistor = resistance × circuit current = 500 × 0.036 = 17.86 V

Example 6.8 A 10 μF capacitor is connected in series with a 2.7 kΩ resistor across a 100 V, 50 Hz supply. Calculate:
  (a) the phase angle between voltage and current;
  (b) the voltage across the capacitor.
What value of inductance would produce the same phase angle?

(a) The phase angle φ may be obtained from $\tan \phi = \dfrac{X_C}{R}$,

where $X_C = \dfrac{1}{2\pi fC}$, i.e. $\tan \phi = \dfrac{1}{2\pi fCR}$

For this circuit, $\tan \phi = \dfrac{10^6}{2\pi \times 50 \times 10 \times 2.7 \times 10^3} = 0.1179$

and φ = 6.72° by calculator or from tables

The phase angle is 6.72°.

(b) To find the voltage across the capacitor the circuit current and the capacitor reactance is needed.

$$X_C = \dfrac{1}{2\pi fC} = \dfrac{10^6}{2\pi \times 50 \times 10} = 318.3 \ \Omega$$

Circuit impedance = $\sqrt{(318.3^2 + 2700^2)} = 2718.7 \ \Omega$

Circuit current = applied voltage/impedance = $\dfrac{100}{2718.7} = 36.78$ mA

Capacitor voltage = reactance × current
 = $318.3 \times 36.78 \times 10^{-3} = 11.71$ V

The voltage across the capacitor is 11.71 V.

Inductive reactance = $2\pi fL$ and this is to equal capacitive reactance to produce the same phase angle.

i.e. $2\pi fL = 318.3$ and $L = \dfrac{318.3}{2\pi \times 50} = 1$ H

Alternatively, $\tan \phi = \dfrac{X_L}{R} = 0.1179$ so that $X_L = 0.1179 \times 2.7 \times 10^3$

and $L = \dfrac{0.1179 \times 2.7 \times 10^3}{2 \times \pi \times 50} = 1$ H

An inductance of 1 H would produce the same phase angle.
(The current would lag the voltage, of course.)

Example 6.9 A variable capacitor is connected in series with a 5 H inductor across a 50 Hz supply. The capacitor is adjusted until the phase angle between voltage and current is zero. Calculate the value of the capacitance when this occurs.

The inductance in the circuit on its own would cause the current to lag the voltage, the capacitance on its own would cause the current to lead the voltage. When combined, the phase angle may be a lead or lag depending upon whether the inductive reactance is greater than or smaller than the capacitive reactance. When the two reactances are equal the phase angle is zero.

Phase angle due to the inductance is given by $\tan\phi = \dfrac{X_L}{R}$

and for the capacitor by $\tan\phi = \dfrac{X_C}{R}$

When these phase angles are equal, $\dfrac{X_L}{R} = \dfrac{X_C}{R}$ and $X_C = X_L$

or $\dfrac{1}{2\pi fC} = 2\pi fL$ so that $C = \dfrac{1}{4\pi^2 f^2 L} = \dfrac{1}{4\pi^2 \times 50^2 \times 5} = 2.03\ \mu F$

The capacitance is 2.03 μF.

Specific objectives

The expected learning outcome is that the student:
8.21 Uses phasor diagrams to solve simple series L, C and R a.c. circuits.
8.22 Defines series resonance as occurring when the supply voltage and current are in phase.
8.23 Sketches a phasor diagram showing that $V_L = V_C$ at series resonance.
8.24 Shows that V_L and V_C may be many times supply voltage.

Series L-C-R circuits

Example 6.9 showed the effect of adding a capacitor in series with an inductor. The combination results in a phase angle which may be leading, lagging or zero depending upon the respective magnitudes of the capacitive reactance and the inductive reactance. Since this in turn depends upon frequency, the phase angle depends upon frequency.

Inductive reactance is directly proportional to frequency, rising as the frequency of the supply is increased. Capacitive reactance is indirectly proportional to frequency and falls as frequency is increased. This is shown in the graphs of X_L plotted against frequency and X_C plotted against frequency in fig. 6.7. At one particular frequency the reactances are equal and the phase angle is zero. At this point *series resonance* is said to occur and the frequency is called the *resonant frequency*, symbol f_r.

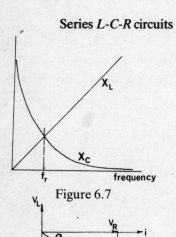

Figure 6.7

Before examining what happens in the circuit at resonance the conditions at frequencies above and below the resonant frequency will be considered. See fig. 6.8.

Conditions below resonance are shown in fig. 6.8a.

Here, because X_L is smaller than X_C, V_L is smaller than V_C

To obtain V_S, an amount equal to V_L is removed from V_C to give the resultant reactive voltage $V_C - V_L$. This is summed with V_R to give V_S *lagging* the current at phase angle ϕ.

Below resonance, $V_S^2 = (V_C - V_L)^2 + V_R^2$ or $Z^2 = (X_C - X_L)^2 + R^2$

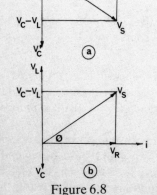

Figure 6.8

(which is obtained by dividing throughout by i^2)

and $Z = \sqrt{[(X_C - X_L)^2 + R^2]}$

Above resonance X_L is greater than X_C and V_L is greater than V_C (Fig. 6.8b).

An amount equal to V_C is removed from V_L to give the resultant reactive voltage $V_L - V_C$. This is summed with V_R to give V_S *leading* the circuit current by phase angle ϕ.

$$\text{Above resonance, } V_S^2 = (V_L - V_C)^2 + V_R^2$$

$$\text{or } Z^2 = (X_L - X_C)^2 + R^2$$

giving $Z = \sqrt{[(X_L - X_C)^2 + R^2]}$

In general $Z = \sqrt{[(X_L \sim X_C)^2 + R^2]}$

where $\sim$ means 'the difference between' given by whichever is the smaller being taken from whichever is the larger. Also from the diagrams

$$\sin\phi = \frac{X_L \sim X_C}{Z}, \quad \cos\phi = \frac{R}{Z} \text{ and } \tan\phi = \frac{X_L \sim X_C}{R}$$

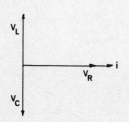

Figure 6.9

Circuit conditions at resonance are shown in fig. 6.9. Here the reactive voltages V_L and V_C are equal and cancel each other. The phase angle is zero and the circuit overall appears resistive.

$$Z = \sqrt{[(X_L \sim X_C)^2 + R^2]} = R^2 = R, \text{ since } X_L = X_C$$

and $\sin\phi = \tan\phi = 0$, and $\cos\phi = 1$, so that ϕ is zero.

The resonant frequency may be obtained from the equation $X_L = X_C$.

$$2\pi f_r L = \frac{1}{2\pi f_r C}, \quad f_r^2 = \frac{1}{4\pi^2 LC} \quad \text{and } f_r = \frac{1}{2\pi\sqrt{(LC)}}$$

At resonance $V_L = V_C$, and since $V_S^2 = (V_L \sim V_C)^2 + V_R^2$

$$V_S^2 = V_R^2 \text{ and } V_S = V_R$$

$$\text{Circuit current } i = \frac{V_S}{Z} = \frac{V_S}{R}$$

$$\text{Hence } V_L = iX_L = \frac{V_S}{R}X_L$$

so that inductive voltage $= \left(\frac{X_L}{R}\right) \times$ supply voltage.

$$\text{Similarly, } V_C = iX_C = \frac{V_S}{R}X_C$$

and capacitor voltage $= \left(\frac{X_C}{R}\right) \times$ supply voltage.

$\frac{X_L}{R}$ and $\frac{X_C}{R}$ are, of course, equal since $X_L = X_C$

The ratio X_L/R or X_C/R may have a value many times greater than unity, for X_L or X_C may be many times greater than R. This means that since the inductive voltage or capacitor voltage is equal to the supply voltage *multiplied by* this ratio, the inductive or capacitor voltage may be many times *larger* than the supply voltage. This characteristic of a series resonant circuit is called *voltage magnification*.

It may be a little difficult to grasp at first, since it appears that a voltage is appearing from nowhere. The reason it is possible is because whatever the size of voltage across the circuit inductance, whether it is many times larger than the supply voltage or not, there is an equal and *opposite* voltage across the capacitor, and overall in the circuit as a whole, the two cancel. Voltage magnification is used in radio receivers to increase the size of the incoming signal.

The ratio $\frac{X_L}{R}$ or $\frac{X_C}{R}$ at resonance is called the voltage magnification or 'Q-factor' of the circuit.

$$Q = \frac{X_L}{R} \text{ or } \frac{X_C}{R} = \frac{\omega L}{R} \text{ or } \frac{1}{\omega CR}$$

Example 6.10 A series circuit consists of a 2 H, 150 Ω inductor in series with a 4 μF capacitor. A variable-frequency 40 V a.c. supply is connected to the circuit and the frequency increased from zero. Determine:

(a) the frequency at which the resultant circuit reactance is zero;
(b) the voltage across the capacitor at this frequency;
(c) the circuit current at this frequency;

(a) The frequency at which the resultant circuit reactance is zero is the resonant frequency.

$$f_r = \frac{1}{2\pi \sqrt{(LC)}} = \frac{1}{2\pi \sqrt{(2 \times 4 \times 10^{-6})}} = 56.27 \text{ Hz}$$

(b) The capacitor voltage at resonance,

$$V_C = \frac{X_C}{R} \times \text{supply voltage}$$

$$= \frac{1}{2\pi f_r CR} \times \text{supply voltage} = \frac{40}{2\pi \times 56.27 \times 4 \times 10^{-6} \times 150}$$

$$= 188.56 \text{ V (nearly five times the supply voltage)}$$

(c) Circuit current at resonance is given by dividing the supply voltage by the *resistance*.

$$\text{Circuit current} = \frac{40}{150} = 0.267 \text{ A}$$

Example 6.11 A coil of inductance 2 mH and resistance 6 Ω forms part of a series resonant circuit, the resonant frequency being 1 MHz. Calculate the capacitance of the capacitor in the circuit and the circuit Q-factor.

The capacitance may be obtained from the equation for the resonant frequency.

$$f_r^2 = \frac{1}{4\pi^2 LC} \quad \text{and } C = \frac{1}{4\pi^2 L f_r^2} = \frac{1}{4\pi^2 \times 2 \times 10^{-3} \times 10^6 \times 10^6}$$

$$= 12.65 \text{ pF}$$

The Q-factor may be obtained from either $Q = \dfrac{X_L}{R}$ or $\dfrac{X_C}{R}$

The equation containing X_L is the better one to use since it involves given and not calculated quantities. Using quantities calculated elsewhere in the problem introduces a risk of error. Sometimes, of course, this is unavoidable.

$$\text{So } Q = \frac{\omega L}{R} = \frac{2\pi f L}{R} = \frac{2\pi \times 10^6 \times 2 \times 10^{-3}}{6} = 2094.4$$

Example 6.12 Calculate the inductance and resistance of a coil which is connected in series with a 0.01 μF capacitor to give a resonant frequency of 500 Hz and a Q-factor of 150.

To obtain the inductance, use $f_r^2 = \dfrac{1}{4\pi^2 LC}$ i.e. $L = \dfrac{1}{f_r^2 4\pi^2 C}$

$$= \frac{10^6}{500^2 \times 4\pi^2 \times 0.01}$$

$$= 10.13 \text{ H}$$

To obtain the coil resistance, use $Q = \dfrac{1}{2\pi f_r CR}$

$$\text{i.e. } R = \frac{1}{2\pi f_r CQ}$$

(again using an equation not involving calculated quantities)

$$R = \frac{10^6}{2\pi \times 500 \times 0.01 \times 150} = 212.2 \ \Omega$$

A 10.13 H, 212.2 Ω coil is required.

Specific objectives

The expected learning outcome is that the student:
8.13 States that power dissipation is I^2R.
8.14 Shows graphically that where currents and voltages are sinusoidal:
 (a) for a purely resistive a.c. circuit, average power is VI.
 (b) for a purely reactive a.c. circuit, average power is zero.
 (c) for a resistive/reactive a.c. circuit, average power depends upon phase angle.
8.15 States that $P = VI \cos \phi$ for sinusoidal waveforms.
8.16 Derives the power triangle from the voltage triangle.
8.17 Identifies true power (P), apparent power (S) and reactive volt-amperes (Q).
8.18 Defines power factor as true power/apparent power and shows that where V and I are sinusoidal, power factor $= \cos \phi$.

8.19 Applies equations in 8.9 and 8.11 to the solution of single-branch L-R series circuits at power and radio frequencies.
8.20 Explains that power dissipation in series L-R and C-R a.c. circuits is I^2R.

Power in a.c. circuits

Power is the rate of doing work or using energy, the unit being the joule per second or *watt*. In electrical circuits the product of voltage and current at any instant gives the rate of using energy, for:

$$\text{Since volt} = \frac{\text{joule}}{\text{coulomb}} \quad \text{and ampere} = \frac{\text{coulomb}}{\text{second}}$$

$$\text{then volt} \times \text{ampere} = \frac{\text{joule}}{\text{coulomb}} \times \frac{\text{coulomb}}{\text{second}} = \frac{\text{joule}}{\text{second}}$$

In d.c. circuits, once steady state values are reached, voltage and current levels are unchanging and the power is calculated easily by multiplying together the voltage and current. In a.c. circuits, even at steady state, the voltages and currents are changing in value from instant to instant and although instantaneous power may be calculated in the same way (instantaneous voltage × instantaneous current), care must be taken in calculating power using particular given values to see that the right kind of value is being used (the r.m.s. value).

The second factor to be taken into consideration is the nature of the circuit; circuits containing reactance behave quite differently from circuits containing resistance alone as is shown below.

Fig. 6.10 shows typical voltage and current waveforms in a circuit containing resistance alone (part a), inductance alone (part b) and capacitance alone (part c). On each graph the waveform of the instantaneous power is also inserted. This waveform is obtained by multiplying together the instantaneous values of voltage and current.

In a purely resistive circuit, voltage and current are in phase and when the voltage reverses its direction of action the current reverses its direction of flow. The product of voltage and current thus always has the same polarity – denoting the region above the time axis as positive, as is usual – the instantaneous power is then always positive. This means, of course, that energy is being used in the circuit all the time. It is in fact converted in the resistor to heat (and possibly light in some applications).

The average power used in the circuit as shown in fig. 6.10a is equal to

$$\frac{V_M I_M}{2}, \text{ i.e. } 0.5 V_M I_M$$

since the power waveform is symmetrical about this value, and this can be written as

$$\sqrt{(0.5)} V_M \times \sqrt{(0.5)} I_M \text{ or } 0.707 V_M \times 0.707 I_M$$

i.e. the r.m.s. value of voltage × r.m.s. value of current.

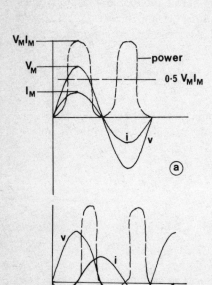
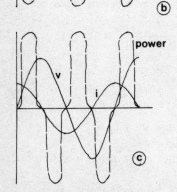

Figure 6.10

So that provided r.m.s. values are used, the power consumed by a purely resistive circuit supplied with alternating current is equal to the product of voltage or current. Denoting r.m.s. values by V and I respectively,

$$\text{Power in resistance} = VI \text{ or } \frac{V^2}{R} \text{ or } I^2R$$

which are the same relationships as in the d.c. case. It must be remembered, however, that r.m.s. values *must* be used.

In figs 6.10b and 6.10c there is a phase difference between voltage and current due to reactance. The instantaneous power curve (obtained by multiplying voltage and current at each instant and *taking into account their polarity*), lies equally displaced about zero and we appear to have 'negative power' for the same time periods as 'positive power'. Negative power periods are those in which current is flowing in a direction not corresponding with the direction of action of the voltage – due to the delays in the circuit caused by reactance.

What is actually happening here is that during the 'positive power' periods energy is taken *from* the supply and stored either in the magnetic field, if the component is inductive ($\frac{1}{2}LI^2$), or in the electric field, if the component is capacitive ($\frac{1}{2}CV^2$). During the 'negative power' periods this energy is returned *to* the supply, the net result being that no power is consumed by the circuit.

The product of voltage and current is thus *not* the power consumed when a circuit is purely reactive. It is given a special name – *reactive volt-amperes*, abbreviated VAr.

When resistance and reactance are combined in an a.c. circuit there is a resultant phase angle between supply voltage and current and the power taken by the circuit must depend in some way upon the value of the phase angle. For a zero phase angle, the power is the voltage–current product and for a $\pi/2$ rad phase angle the power is zero. When the phase angle lies between zero and $\pi/2$ rad the power consumed lies between the supply voltage–current product and zero. It is in fact multiplied by a constant which is determined by the circuit.

This constant for a particular circuit at a particular frequency is called the *power factor* of the circuit.

$$\text{Power} = V_S I \times \text{power factor}$$

where V_S and I are the r.m.s. values of the supply voltage and current. For a purely resistive circuit the power factor is unity (one), and for a purely reactive circuit the power factor is zero.

To find out more about the power factor and how it is affected by the current consider fig. 6.11, which shows how what is called the power triangle of a resistive–reactive circuit is constructed. The circuit considered is an *L-R* circuit but the theory applies equally to a *C-R* circuit or an *L-R* circuit. In all cases there is a resistive voltage and a reactive voltage summed to give the supply voltage.

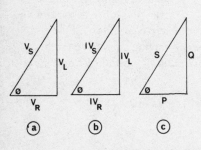

Figure 6.11

For this circuit part (a) of the figure shows the resistive voltage V_R, the reactive voltage V_L and the supply voltage V_S. If each side of the triangle is multiplied by the circuit current I, the triangle of fig. 6.11b is obtained with sides IV_S, IV_R and IV_L.

Now IV_R is the actual power absorbed by the circuit (since only resistance absorbs power), IV_L are the reactive volt-amperes and IV_S, the product of the current and supply voltage, is called the *apparent power* of the circuit. Apparent power is measured in volt-amperes (VA) *not* watts. The symbol for actual power is P, for reactive volt-amperes is Q and for apparent power is S. The triangle is redrawn showing these symbols in fig. 6.11c.

The angle between V_S and V_R is the phase angle between V_S and I (since I and V_R are always in phase) and is shown as ϕ. This must also be the angle between IV_S and IV_R and thus between S, the apparent power, and P, the actual power.

$$\text{From the triangle, } \cos\phi = \frac{P}{S} \text{ and } P = S\cos\phi$$

i.e. actual power = apparent power × $\cos\phi = IV_S \cos\phi$

$\cos\phi$ is the power factor of the circuit

Another way of looking at the situation is that the current I may be considered to have two components $I\cos\phi$ in phase with V_R, which when multiplied by V_R gives the true power, and $I\sin\phi$ at a phase angle of $\pi/2$ rad with V_R, which when multiplied by V_L gives the reactive volt-amperes, for, from the triangle again,

$$\sin\phi = \frac{Q}{S} \text{ and } Q = S\sin\phi = IV_S \sin\phi \text{ or } V_S \times I\sin\phi$$

$$P = IV_S \cos\phi \text{ or } V_S \times I\cos\phi$$

The power factor of a resistive–reactive circuit is the ratio between actual and apparent power. It is equal to the cosine of the phase angle between supply voltage and supply current.

It should be noted that all the foregoing theory and definitions apply to voltages and currents having *sinusoidal waveforms* since these have been used throughout in waveform diagrams to obtain the various phase angles in resistive and reactive circuits.

To summarise: For any series circuit having a resistive voltage V_R and a reactive voltage V_L (or V_C), a supply voltage V_S and supply current I and phase angle ϕ

$$\text{True power } P = IV_R \text{ or } I^2R \text{ or } \frac{V_R^2}{R}$$

$$\text{Apparent power } S = IV_S$$

$$\text{Reactive volt-amperes } Q = IV_L \text{ (or } IV_C\text{)}$$

$$\text{Power factor} = \frac{\text{true power}}{\text{apparent power}} = \cos\phi$$

$$\text{Also } P = S\cos\phi \text{ and } Q = S\sin\phi$$

Example 6.13 A series circuit consists of a 10 H, 200 Ω inductor in series with a 200 Ω resistor. The supply voltage is 240 V, 50 Hz. Calculate:
(a) the power factor of the circuit;
(b) the true power, apparent power and reactive volt-amperes.

(a) The phase angle is given by $\tan\phi = \dfrac{X_L}{R} = \dfrac{2\pi f L}{R}$

$$= \dfrac{2\pi \times 50 \times 10}{400} = 7.854$$

(note the total resistance is 200 + 200 i.e. 400 Ω)

and $\phi = 82.74°$ so that $\cos\phi = 0.1263$

The power factor is 0.1263

(b) To find these answers we first of all require the circuit current. This is obtained by dividing the supply voltage by circuit impedance.

Circuit impedance $= \sqrt{(X_L^2 + R^2)} = \sqrt{[(2\pi f L)^2 + R^2]}$
$= \sqrt{[(2\pi \times 50 \times 10)^2 + (400)^2]} = 3166.9$ Ω

Circuit current = 240/3166.9 = 0.07578 A

Apparent power = 240 × 0.07578 = 18.18 VA (*not* watts)

True power = 18.18 × $\cos\phi$ = 18.18 × 0.1263 = 2.297 W

Reactive volt-amperes = 18.18 × $\sin\phi$ = 18.18 × sin 82.74°
= 18.04 VAr

True power is 2.297 W, apparent power 18.18 VA (note the unit) and reactive volt-amperes 18.04 VAr.

Example 6.14 A 2 H, 50 Ω inductor is connected in series with a 0.4 μF capacitor across a 100 Hz a.c. supply. Determine the circuit power factor.

To find the power factor we need the *resultant* phase angle. This in turn requires the determination of the resultant reactance.

$$X_L = 2\pi f L = 2\pi \times 100 \times 2 = 1256.6 \text{ Ω}$$

$$X_C = \dfrac{1}{2\pi f C} = \dfrac{10^6}{2\pi \times 100 \times 0.4} = 3978.9 \text{ Ω}$$

The resultant reactance = 3978.9 − 1256.6

Thus $X_C - X_L = 2722.3$ Ω

$$\tan\phi = \dfrac{X_C - X_L}{R} = \dfrac{2722.3}{50} = 54.44$$

$\phi = 88.94°$ and power factor, $\cos\phi = 88.94° = 0.018$

The power factor is 0.018

Example 6.15 An inductor is connected across a 240 V, 50 Hz supply and the power as measured by a wattmeter is 15 W. The circuit current is 0.3 A. Calculate the inductance and resistance of the inductor.

Apparent power = 240 × 0.3 = 72 VA; true power = 15 W

$$\text{Power factor, } \cos\phi = \frac{15}{72} = 0.208$$

Hence $\phi = 77.97°$ and $\tan\phi = 4.69$

$$\text{Now } \tan\phi = \frac{X_L}{R}, \text{ so that } \frac{X_L}{R} = 4.69$$

or $X_L = 4.69R$

which gives us one equation connecting X_L and R. For two unknowns we require two equations; the other may be obtained from the impedance equation:

Since $Z^2 = X_L^2 + R^2$ and $Z = \text{voltage/current} = \dfrac{240}{0.3} = 800\ \Omega$

then $800^2 = X_L^2 + R^2$. But $X_L = 4.69R$,

so $800^2 = (4.69R)^2 + R^2 = 22R^2 + R^2 = 23R^2$

and $R = \sqrt{\left(\dfrac{800^2}{23}\right)} = 166.8\ \Omega$

$X_L = 4.69R = 4.69 \times 166.8 = 782.35\ \Omega$

i.e. $2\pi fL = 782.35$, so that $L = \dfrac{782.35}{2\pi \times 50} = 2.49$ H

The resistance is 166.8 Ω, the inductance is 2.49 H.

Summary

Resistance has the same effect on alternating current as on direct current. Ohm's Law, $V = IR$, still applies and there is no phase shift between voltage and current. In using Ohm's Law care must be taken to use the same kind of value (r.m.s., peak, etc.).

The opposition to alternating current due to inductance alone is called inductive reactance, X_L. It is obtained by dividing the inductive voltage by the current and is measured in ohms. Inductive reactance causes the applied voltage to lead the resultant current by $\pi/2$ radians. For an inductance L henrys and supply frequency f hertz

$$X_L = 2\pi fL$$

The opposition to alternating current due to capacitance alone is called capacitive reactance, X_C. It is obtained by dividing the capacitor voltage by the current and is measured in ohms. Capacitive reactance causes the applied voltage to lag the resultant current by $\pi/2$ radians. For a capacitance C farads and supply frequency f hertz,

$$X_C = \frac{1}{2\pi fC}$$

The opposition to alternating current due to a circuit containing resistance and either inductance or capacitance or both is called impedance, Z. It is obtained by dividing the supply voltage by current and is measured in ohms. The phase shift between voltage and current depends upon the resultant reactance and resistance of the circuit and lies between $\pi/2$ radians, voltage lagging current, and $\pi/2$ radians, voltage leading current.

For an inductive–resistive circuit the impedance is given by

$$Z^2 = R^2 + X_L^2 \text{ and the phase angle } \phi \text{ by}$$

$$\sin\phi = \frac{X_L}{Z} \text{ or } \cos\phi = \frac{R}{Z} \text{ or } \tan\phi = \frac{X_L}{R}$$

For a capacitive–resistive circuit the impedance is given by

$$Z^2 = R^2 + X_C^2 \text{ and the phase angle } \phi \text{ by}$$

$$\sin\phi = \frac{X_C}{Z} \text{ or } \cos\phi = \frac{R}{Z} \text{ or } \tan\phi = \frac{X_C}{R}$$

For a circuit containing resistance, inductance and capacitance the impedance Z is given by

$$Z^2 = R^2 + (X_L \sim X_C)^2, \text{ where the symbol } \sim \text{ means 'the difference between'.}$$

The phase angle ϕ is given by $\sin\phi = \dfrac{X_L \sim X_C}{Z}$ or $\cos\phi = \dfrac{R}{Z}$

$$\text{or } \tan\phi = \frac{X_L \sim X_C}{R}$$

In a circuit containing resistance, inductance and capacitance there is one frequency when $X_L = X_C$. This is called the resonant frequency, f_r,

$$\text{and since } 2\pi f_r L = \frac{1}{2\pi f_r C}, \; f_r = \frac{1}{2\pi \sqrt{(LC)}}$$

The circuit is said to be resonant or at resonance at this frequency. The impedance at the resonant frequency is equal to the circuit resistance since the reactances are equal and opposite.

At resonance the voltage across the inductance (which is equal to the voltage across the capacitance) may be many times larger than the supply voltage and voltage magnification is said to occur.

At the resonant frequency the supply voltage V_S is given by $V_S = i_S R$ where i_S is the supply current. Also,

$$V_L = i_S X_L \text{ and } V_C = i_S X_C$$

where V_L and V_C represent the voltages across the inductance and capacitance respectively.

The ratio between reactive voltage and supply voltage, $\dfrac{V_L}{V_S}$ or $\dfrac{V_C}{V_S}$

$$\text{is given by } \frac{V_L}{V_S} = \frac{i_S X_L}{i_S R}$$

$$\text{or } \frac{V_C}{V_S} = \frac{i_S X_C}{i_S R}, \text{ i.e. } X_L/R \text{ or } X_C/R$$

This ratio is called the voltage magnification or Q-factor of the circuit. It has the symbol Q.

$$Q = \frac{X_L}{R}, \text{ i.e. } \frac{2\pi fL}{R},$$

$$\text{or } \frac{X_C}{R}, \text{ i.e. } \frac{1}{2\pi fCR}$$

The apparent power in an a.c. circuit, symbol S, is the product of the r.m.s. values of voltage and current. Denoting these by V and I, $S = VI$.

Apparent power is measured in volt-amperes, VA. The true power in an a.c. circuit, symbol P, is the product of the apparent power and the cosine of the phase angle between voltage and current. Denoting phase angle by ϕ,

$$P = S \cos \phi = VI \cos \phi$$

True power is measured in watts, W.

Cos ϕ is called the circuit power factor and its value lies between zero (a totally reactive circuit) and unity (a totally resistive) circuit).

The reactive volt-amperes in an a.c. circuit, symbol Q, is the product of the apparent power and the sine of the circuit phase angle

$$Q = S \sin \phi = VI \sin \phi$$

It is measured in reactive volt-amperes, VAr.

EXERCISE 6

1. Calculate the peak current flowing in a 150 Ω resistor placed across a 240 V, 50 Hz supply.

2. Calculate the inductive reactance and impedance of a 10 H, 200 Ω inductor at 150 Hz.

3. A 0.4 μF capacitor is placed in series with a resistor across a 240 V, 50 Hz supply. The circuit current is 24 mA. Calculate the value of the resistance.

4. Calculate the impedance of a circuit containing a 14 H, 50 Ω choke in series with a 0.01 μF capacitor at 150 Hz.

5. The reactance of a 5 H inductor is equal to the reactance of a certain capacitor at a frequency of 0.75 kHz. What is the value of the capacitor?

6. The current flowing in a series C-R circuit placed across an alternating current supply of frequency 125 Hz is 0.25 A. The resistor voltage is 115 V, the voltage across the capacitor is 190 V. Calculate the supply voltage, the circuit impedance and the values of the capacitance and resistance.

7. The impedance of an inductor at 500 Hz is 500 Ω. When a direct voltage of 200 V is applied the steady state current is 0.5 A. Calculate the inductance and resistance of the inductor.

8. When 200 V, 150 Hz is applied to a resistor and capacitor in turn, the current flowing is 0.02 A and 0.15 A respectively. Calculate the current flowing in a circuit containing these components connected in series when the same voltage is applied to the circuit as a whole.

9. A 10 H, 150 Ω inductive coil is connected in series with a 0.02 μF capacitor across a 50 Hz a.c. supply. When a d.c. supply at the same voltage is applied to the coil alone the steady current is 0.2 A. Calculate the alternating current which flows in the series circuit.

10. Calculate the impedance of a 0.5 H, 75 Ω coil connected in series with a 0.05 μF capacitor across a 100 Hz a.c. supply.

11. When a 240 V, 50 Hz supply is connected across a circuit made up of a capacitor in series with a resistor, the voltage across the capacitor is 75 V. Calculate the voltage across the resistor.

12. Calculate the value of capacitance of a capacitor placed in series with a 5 H, 100 Ω choke across a 150 Hz a.c. supply, if the phase angle between supply voltage and current is

(a) $\frac{\pi}{4}$ rad, current lagging; (b) $\frac{\pi}{4}$ rad, current leading; (c) zero

13. A 0.2 H, 40 Ω coil is placed in series with a capacitor across a 200 V, 100 Hz supply. The resultant phase angle between supply voltage and current is zero. Calculate the:

(a) capacitance of the capacitor;
(b) voltage across the capacitor;
(c) circuit current;
(d) voltage across the inductor.

14. Calculate the resonant frequency and the Q-factor of a 0.2 H, 75 Ω coil in series with a 0.2 μF capacitor.

15. A series L-C-R circuit has a resonant frequency of 500 kHz and a Q-factor of 1500. The coil inductance is 0.01 H. Calculate the coil resistance and the capacitance of the capacitor.

16. The resonant frequency of a circuit made up of a 5 H, 150 Ω inductor in series with a capacitor is 150 Hz. Calculate the phase angle of the circuit at 300 Hz.

17. A 1 H, 150 Ω coil is connected in series with a 330 Ω resistor. Calculate the power factor of the circuit when the supply frequency is 500 Hz. ·

18. A 0.5 μF capacitor is connected in series with a 200 Ω resistor across a 240 V, 50 Hz supply. Calculate:

(a) the circuit power factor;
(b) the true power, apparent power and reactive volt-amperes.

19. An a.c. supply of 500 V at 50 Hz is connected to a 10 H, 200 Ω choke. Calculate the value of the capacitor which when connected in series with the choke reduces the circuit power factor by half. Determine the true power in the circuit before and after the capacitor is connected.

20. The true power absorbed by a certain inductor connected to a 150 V, 100 Hz supply is 12.5 W when the circuit current is 0.1 A. Calculate the inductance and resistance of the inductor and the circuit power factor.

21. The true power absorbed by a reactive circuit is 750 W, the reactive volt-amperes being 800 VAr. Calculate the apparent power and the circuit power factor.

22. The apparent power absorbed by a reactive circuit when connected to a 240 V, 50 Hz supply is 240 VA. The circuit power factor is 0.9. Calculate:

(a) the circuit current; (b) the true power;
(c) the circuit resistance; (d) the circuit reactance.

23. An inductive circuit draws 0.5 A from a 100 V, 50 Hz supply, the true power taken from the supply being 10 W. Calculate the circuit impedance, resistance and inductance.

24. A 5 H, 200 Ω coil is connected in series with a 0.1 μF capacitor across a 240 V variable-frequency supply. Calculate the true power, apparent power and reactive volt-amperes at the resonant frequency of the circuit and at a frequency equal to twice the resonant frequency.

SELF-ASSESSMENT EXERCISE 6

Possible marks

1. Write down the equation relating inductive reactance X_L, frequency f, and inductance L, and the equation relating capacitive reactance X_C, frequency f, and capacitance C. (3)

2. State the relationship between the impedance Z of a circuit and its resistance R, inductive reactance X_L and capacitive reactance X_C. (3)

3. State the relationship between resonant frequency f_r, inductance L, and capacitance C. (3)

4. State the relationship between the Q-factor of a circuit frequency f, inductance L, and resistance R, and the relationship between the Q-factor, frequency f, capacitance C, and resistance R. (3)

5. State the equations for true power, apparent power and reactive volt-amperes in terms of voltage V, current I, and the circuit phase angle ϕ. (3)

6. The Q-factor of a series L-C-R circuit at a frequency of 1 kHz is 100. Calculate the inductance if the circuit resistance is 75 Ω. (5)

7. The phase angle of a series L-R circuit is $\pi/3$ radians. The true power is 500 W when the supply voltage is 240 V. Calculate the circuit current. (5)

8. Calculate the power factor of an a.c. circuit in which the apparent power is 850 VA and the reactive volt-amperes are 425 VAr. (5)

9. Two resistors of equal value are connected in parallel, the combination then being connected in series with a 10 H, 200 Ω coil across a 240 V, 50 Hz supply. The resultant phase angle between supply voltage and current is found to be $\pi/4$ rad. Calculate:
   (a) the value of each of the parallel-connected resistors;
   (b) the true power in the circuit;
   (c) the power consumed in the coil alone. (14)

10. A resonant circuit consists of an inductor and a 0.5 μF capacitor connected in series. The resonant frequency is 500 Hz and at this frequency the voltage across the capacitor is 95 times the supply voltage. Calculate:
    (a) the inductance and resistance of the inductor;
    (b) the phase angle of the inductor alone;
    (c) the frequency at which the capacitive reactance is equal to twice the inductive reactance. (14)

11. The true power absorbed by a series C-R circuit drawing 2 A from a 240 V, 50 Hz supply is 150 W. Calculate the:
    (a) apparent power;
    (b) power factor;
    (c) reactive volt-amperes;
    (d) values of capacitance and resistance;
    (e) value of inductance to obtain resonance at this frequency. (14)

12. The Q-factor of a series resonant circuit containing a 0.02 μF capacitor is 250. At resonance the circuit draws 0.12 A from a 200 V supply. Calculate the:
    (a) circuit resistance;
    (b) resonant frequency
If the frequency of the supply is now doubled calculate the power factor of the circuit at the new value of frequency. (14)

13. A 2 H, 150 Ω coil is connected to a 10 μF capacitor across a 200 V, 100 Hz supply. Calculate the:
    (a) impedance of the circuit;
    (b) resultant reactance of the circuit;
    (c) power factor;
    (d) apparent power;
    (e) reactive volt-amperes (14)

Answers

EXERCISE 6

1. 2.26 A
2. 9424.78 Ω; 9426.9 Ω
3. 6055.9 Ω
4. 92.9 kΩ
5. 9 nF
6. 222.1 V; 888.37 Ω; 460 Ω; 1.67 μF
7. 95.49 mH; 400 Ω
8. 19.82 mA
9. 0.192 mA
10. 31.52 kΩ
11. 227.98 V
12. (a) 0.23 μF (b) 0.221 μF (c) 0.225 μF
13. (a) 12.66 μF (b) 628.3 V (c) 5 A (d) 628.3 V
14. 795.8 Hz; 13.33
15. 20.94 Ω; 10.13 pF
16. 1.55 rad
17. 0.15
18. (a) 0.0314 (b) 9.04 VA; 0.2839 W; 9.038 VAr
19. 1 μF; 5.04 W; 38.92 W
20. 1.3196 H; 1250 Ω; 0.88
21. 1096.6 VA; 0.684
22. (a) 1A (b) 216 W (c) 216 Ω (d) 104.6 Ω
23. 200 Ω; 40 Ω; 0.62 H
24. 288 W, 288 VA, zero VAr; 0.0576 W, 4.07 VA, 4.069 VAr

Marks

SELF-ASSESSMENT EXERCISE 6

1. $X_L = 2\pi fL$; $X_C = 1/2\pi fC$ (1½ each)
2. $Z^2 = R^2 + (X_L \sim X_C)^2$ (3)
3. $f_r = \dfrac{1}{2\pi \sqrt{(LC)}}$ (3)
4. $Q = \dfrac{2\pi fL}{R}$ and $Q = \dfrac{1}{2\pi fCR}$ (1½ each)
5. True power = $VI \cos \phi$; apparent power = VI; reactive volt-amperes = $VI \sin \phi$ (1 each)

6. $Q = \dfrac{2\pi fL}{R}$, i.e., $100 = \dfrac{2\pi \times 1000 \times L}{75}$ (2)

$L = \dfrac{7500}{2000\pi} = 1.19$ H (3)

7. True power $= VI \cos \phi$, i.e., $500 = 240 \times I \times \cos(\pi/3)$ (2)

$I = 4.17$ A (3)

8. Reactive volt-amperes $=$ apparent power $\times \sin \phi$ (2)

$425 = 850 \sin \phi$ and $\sin \phi = \dfrac{425}{850}$

$\phi = 0.52$ rad (2)

Power factor, $\cos \phi = 0.866$ (1)

9. (a) $\tan \phi = \dfrac{X_L}{R}$

so that $R = \dfrac{X_L}{\tan \phi} = \dfrac{2\pi \times 50 \times 10}{1}$ ($\phi = \pi/4$ rad) $= 3141.6 \, \Omega$

Resistance of parallel combination $= 3141.6 - 200 = 2941.6$ (2)

Resistance of each resistor $= 0.5 \times 2941.6 = 1470.8 \, \Omega$ (2)

(b) $Z^2 = R^2 + X_L^2 = (3141.6)^2 + (2\pi \times 50 \times 10)^2$

$Z = 4442.9 \, \Omega$ (2)

Current $I = 240/4442.9 = 0.054$ A (2)

Power factor, $\cos \pi/4 = 0.707$ (2)

True power $P = VI \cos \phi = 240 \times 0.054 \times 0.707 = 9.17$ W (2)

(c) Power consumed in coil $= I^2 \times$ coil resistance $= 0.054^2 \times 200 = 0.58$ W (2)

10. (a) Resonant frequency $f_r = \dfrac{1}{2\pi \sqrt{(LC)}}$ and $f_r^2 = \dfrac{1}{4\pi^2 LC}$

so that coil inductance $L = \dfrac{1}{4\pi^2 \times f_r^2 C} = \dfrac{10^6}{4\pi^2 \times 500^2 \times 0.5} = 0.203$ H (3)

$Q = \dfrac{2\pi fL}{R}$, so that $R = \dfrac{2\pi fL}{Q}$. $Q = 95$, so

coil resistance $R = \dfrac{2\pi \times 500 \times 0.203}{95} = 6.7 \, \Omega$ (3)

(b) If inductor phase angle is represented by θ

$\tan \theta = \dfrac{X_L}{R} = \dfrac{2\pi \times 500 \times 0.203}{6.7} = 95.2$

Hence, inductor phase angle is 1.57 rad. (4)

(c) $X_L = 2\pi fL$ and $X_C = \dfrac{1}{2\pi fC}$

When $X_C = 2X_L$, $\dfrac{1}{2\pi fC} = 4\pi fL$

and $f^2 = \dfrac{1}{8\pi^2 LC} = 1/(8\pi^2 \times 0.203 \times 0.5 \times 10^{-6})$

$f = 353.24$ Hz

Frequency when $X_C = 2X_L$ is 353.24 Hz (4)

11. (a) Apparent power $= VI = 240 \times 2 = 480$ VA (2)

(b) Power factor $= \dfrac{\text{true power}}{\text{apparent power}} = \dfrac{150}{480} = 0.3125$ (3)

(c) Reactive volt-amperes = apparent power $\times \sin \phi$ (where ϕ is the phase angle and $\cos \phi$ the power factor)

$\cos \phi = 0.3125$, thus $\phi = 1.253$ rad and $\sin \phi = 0.9499$.

Reactive volt-amperes $= 480 \times 0.9499 = 455.96$ VAr (3)

(d) True power $= I^2 R$

and resistance $R = $ true power $/I^2 = \dfrac{150}{4} = 37.5 \ \Omega$ (3)

$\text{Tan } \phi = \dfrac{X_C}{R} = \dfrac{1}{2\pi f C R}$

and $C = \dfrac{1}{2\pi f R \tan \phi} = \dfrac{1}{2\pi \times 50 \times 37.5 \times 3.04} = 27.92 \ \mu\text{F}$

($\phi = 1.253$ rad, $\tan \phi = 3.04$) (3)

12. (a) Supply current at resonance, $I = \dfrac{V}{R}$

and circuit resistance $R = \dfrac{V}{I} = \dfrac{200}{0.12} = 1666.67 \ \Omega$ (5)

(b) $Q = \dfrac{1}{2\pi f_r C R}$

so that $f_r = \dfrac{1}{2\pi Q C R} = \dfrac{10^6}{2\pi \times 250 \times 0.02 \times 1666.67}$

Resonant frequency $= 19.1$ Hz (5)

At resonance, $X_L = X_C = \dfrac{10^6}{2\pi \times 19.1 \times 2} = 4166.4 \ \Omega$

At twice the resonant frequency,

X_L has doubled and is equal to 2×4166.4, i.e. $8332.8 \ \Omega$; X_C has halved and is equal to 0.5×4166.4, i.e. $2083.2 \ \Omega$.

If ϕ is the phase angle, $\tan \phi = \dfrac{X_L - X_C}{R} = 6249.6/1666.67$

and $\phi = 1.31$ rad

Thus circuit power factor, $\cos \phi = 0.2577$ (4)

13. (a) $X_L = 2\pi f L = 2\pi \times 100 \times 2 = 1256.6 \ \Omega$

$X_C = \dfrac{1}{2\pi f C} = \dfrac{10^6}{2\pi \times 100 \times 10} = 159.1 \ \Omega$

$X_L - X_C = 1097.5$

$Z^2 = R^2 + (X_L - X_C)^2 = 150^2 + 1097.5^2$ (4)

Impedance $Z = 1107.7 \ \Omega$

(b) Resultant reactance $= X_L - X_C = 1097.5$ from (a) above (1)

(c) $\text{Tan } \phi = \dfrac{X_L - X_C}{R} = \dfrac{1097.5}{150}$ and $\phi = 1.43$ rad

Power factor, $\cos \phi = 0.135$ (3)

(d) Apparent power $= VI = \dfrac{V^2}{Z} = \dfrac{200^2}{1107.7} = 36.11$ VA (3)

(e) Reactive volt-amperes = apparent power $\times \sin \phi$; $\sin \phi = 0.99$

Reactive volt-amperes $= 36.11 \times 0.99 = 35.75$ VAr (3)

7 Semiconductor diodes and transistors

Topic areas: G and H

General objective

The expected learning outome is that the student:
Understands the principles of p- and n-type semiconductor materials. Understands the basic principle of the bipolar p-n-p and n-p-n transistors.

Specific objectives

The expected learning outcome is that the student:
9.1 States the order of resistivities of conductors, insulators and semiconductors.
9.2 Sketches curves of resistance to base of temperature, for conductors, insulators and semiconductors.
9.3 States that silicon and germanium are semiconductors.
9.4 Identifies the different allowable working temperatures for silicon and germanium.
9.5 Explains doping in terms of the inclusion of atoms, in the semiconductor array of bonded atoms, which either donate charge carriers or create 'holes' which can accept charge carriers.
9.6 Sketches curves of current to a base of voltage with forward and reverse bias for a junction.
9.7 Explains the influence of minority carriers in reverse biased junctions.
9.8 Draws the circuit symbol for a p-n junction indicating conventional and electron current flow paths.

In this chapter we shall be concerned with the theory of semiconductors, semiconductor diodes and transistors. Diodes and transistors are called *active* electronic components (unlike resistors, inductors and capacitors which are *passive* components) and are used to control, shape and process electronic signals. The essential characteristic of a diode, which is a two-electrode device, is that it conducts current better in one direction than in the other. The essential characteristic of a transistor is that a small current may be used to control a much larger current and the device may be used to amplify (make larger) electronic signals.

Properties of semiconductors

The heart of modern electronics is the semiconductor. A semiconductor material has a value of resistivity lying between that of a conductor and that of an insulator. Typically, the resistivity would be about 10^7 times that of a conductor and 10^{-13} times that of an insulator or, putting it another way, a conductor has a conductivity about ten million times that of a semiconductor; a semiconductor, in turn, has a conductivity about ten billion (million million) times that of an insulator. A second important characteristic is that a semiconductor has a negative temperature coefficient of resistance which means that as the temperature of the material rises its resistance falls. Typical resistance–temperature curves are shown in fig. 7.1.

As shown in the figure, the resistance of a conductor rises with temperature but the resistance of semiconductors and insulators falls. When the temperature of any material rises, the internal energy of the material, that is, the energy of the atoms and electrons within the material, rises. The rise in temperature has two effects: firstly, electrons gain sufficient energy to leave the bonds binding them to the parent atoms and secondly, the vibration of the atoms within the material due to their energy increases.

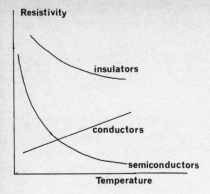

Figure 7.1

The first effect releases more electrons for conduction, should a p.d. be applied across the material. The second effect increases the difficulty encountered by the electrons in moving through the material. The first effect tends to reduce resistance, the second to increase it. Which is the greater depends upon the atomic structure of the material.

A conductor is a material which has a relatively large number of loosely bound electrons in any case and the further release of electrons has less effect than the increased movement of the atoms so that resistance rises. With semiconductors and insulators, which have correspondingly fewer electrons available for conduction, the increase in the number due to a temperature rise has a relatively much greater effect and the resistance falls.

Typical semiconductor materials used in electronics are germanium and silicon, silicon being by far the most common. Both germanium and silicon are *crystalline* in nature, that is, they exist in the form of crystals. The main characteristic of a crystal is its definite geometrical shape formed by flat faces, edges and corners, with specific angles between the faces and edges. Common salt and graphite are two well-known crystalline materials; they are not, however, semiconductors.

The atomic structure of a crystal, that is, the way in which the atoms are interconnected, causes it normally to be a very stable substance without the free or loosely bound electrons characteristic of a good conductor. However, when the material absorbs energy due to the temperature of the surroundings, electrons may leave their bonds within the semiconductor crystal and it is possible to set up an electric current. To understand this we must take a closer look at the bonds within a semiconductor.

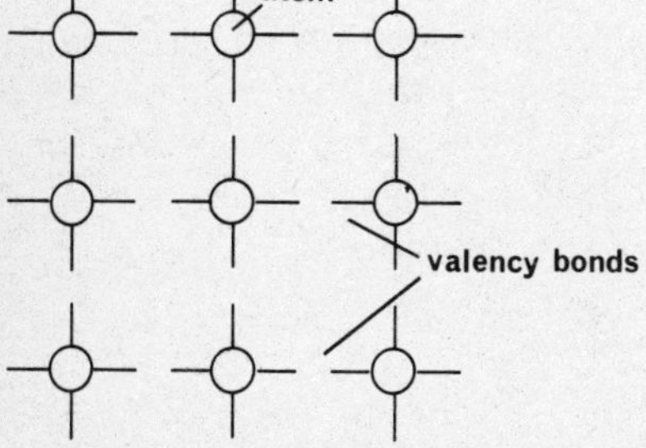

Pure silicon or germanium is *tetravalent*. This means it has four (tetra) electrons per atom available for linking other neighbouring atoms. Such electrons are called *valency* electrons. A way of showing this in diagram form is given in fig. 7.2. The valency electrons interlink the atoms forming a strong bond between the atoms. When the crystal temperature rises as heat energy is absorbed from the surroundings, a few of the valency electrons may leave these bonds and if a p.d. is applied across the crystal a small electric current may flow. Thus pure silicon or germanium will conduct slightly at usual

Figure 7.2

Figure 7.3

Figure 7.4

room temperatures. Another name for 'pure' in this particular usage is *intrinsic*. An intrinsic semiconductor crystal has no impurities within it.

If, however, very small controlled amounts of certain impurities are added to the intrinsic material so that it is no longer pure, the conducting properties of the crystal can be changed. It is this addition of small, controlled amounts of impurity that alters the semiconductor properties and allow us to use it in the semiconductor diode and other semiconductor devices such as transistors.

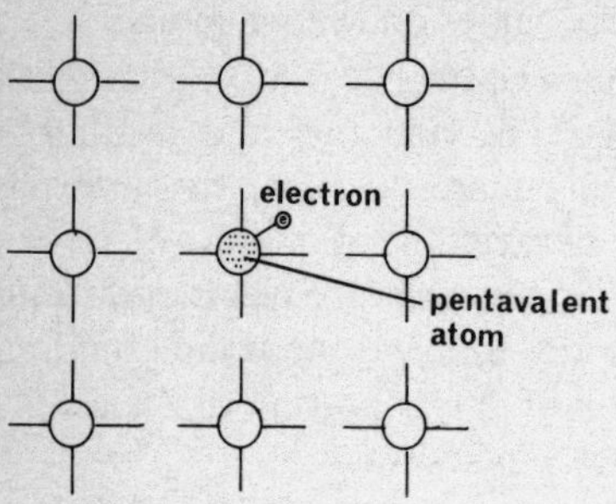

Look again at fig. 7.2. Each of the four valency electrons in each atom link with adjacent atoms to form the strong bond typical of an intrinsic semiconductor crystal. If now a pentavalent material, such as arsenic, having *five* valency electrons per atom is introduced, as shown in fig. 7.3, four of the arsenic atom valency electrons link up with adjacent semiconductor atoms, forming the strong bond again, but the fifth arsenic valency electron has no atom with which to link and remains bound only to its 'parent' atom. By comparison to the other electrons it is relatively loosely bound and if a p.d. is applied to this crystal a much larger current flows than if the same p.d. is applied across the same size of intrinsic crystal.

This crystal is said to be *doped* with impurity and is called an *extrinsic* crystal. The addition of the impurity has increased the number of charge carriers, in this case electrons, and, since electrons are negatively charged, the material is called *n-type* semiconductor. It should be noted, however, that n-type semiconductor is not charged negatively overall since each atom, whether it is semiconductor or impurity, is electrically neutral and therefore so is the overall crystal.

It is also important to realise that the doped crystal conducts equally well in either direction. It does not on its own have the conduction properties of a diode, which conducts better in one direction than in the other.

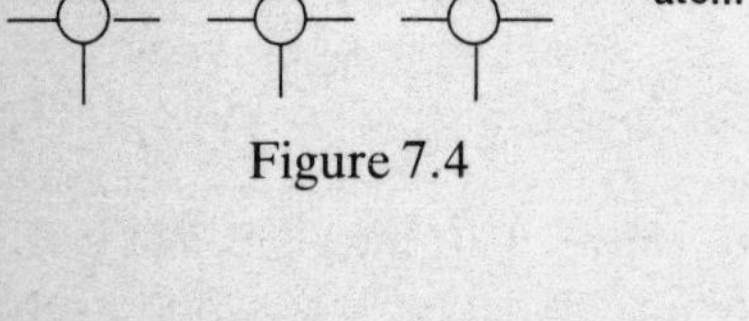

Addition of a pentavalent impurity to a semiconductor crystal increases the available number of negative charge carriers and changes the conductive properties of the material. These properties may also be changed by adding to the intrinsic material very small quantities of an impurity which is *trivalent*. A trivalent material has *three* valency electrons and typical impurities added to pure silicon or germanium are indium and gallium.

When a trivalent impurity atom is added to the crystal we have the situation shown in fig. 7.4. The three valency electrons of the impurity bond with three neighbouring atoms, but the bond between the fourth neighbouring atom and the impurity atom is incomplete. Instead of containing two electrons, one valency electron from each of the bonded atoms, it contains only one. This electron is from the semiconductor atom, there being no further valency electrons available from the impurity atom. The resulting incomplete bond has a vacancy or 'hole' requiring an electron for the bond to be complete.

The existence of these holes within the extrinsic crystal changes the conductive properties from those of the pure crystal because the incomplete bond tends to attract other electrons in the vicinity. The stability of the pure crystal which has all its bonds complete is thus weakened. What now happens is that an electron from a neighbouring two-electron bond may leave this bond and fill the hole in the incomplete bond. The incomplete bond is now complete, but in order for this to be so, what was a neighbouring complete bond is now incomplete. The hole has moved.

Under the influence of a p.d. applied across the crystal, this hole movement is increased and what are normally valency electrons move through the crystal. The doped crystal conductivity thus rises to a value higher than that of the pure crystal. Again, current flow occurs in either direction through the crystal, determined by which way round the p.d. is applied and no diode conducting properties are present. In a sense, the hole acts similarly to a positive charge, although it should be emphasised that it is not, and because of this the doped crystal is called *p-type* semiconductor. As with the n-type semiconductor, the doped crystal is not electrically charged overall since each atom, semiconductor or impurity, is electrically neutral.

Diodes

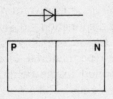

Figure 7.5

Neither p-type nor n-type semiconductor crystals have conductive properties which are different in one direction from the other. In both types of semiconductor the conductivity is higher than in the intrinsic semiconductor but it is not dependent on direction of application of a p.d. However, when p-type and n-type semiconductors are formed within the *same* crystal, such that the p-type material lies adjacent to and alongside the n-type material, it is found that conduction in one direction far exeeds the conduction in the other direction when the same p.d. is applied in each direction in turn. This arrangement, shown in diagram form in fig. 7.5, is that of the semiconductor diode.

In any material at temperatures above that of absolute zero there is absorption of energy by electrons and if they are able to do so they move about within the material. The movement is quite random, taking place in all directions, and is dependent on the amount of energy absorbed.

For the arrangement shown in fig. 7.5 both p-type and n-type materials are initially electrically neutral and movement of electrons occurs in both materials. In the p-type material, valency electrons move from hole to hole and in the n-type material, the loosely-bound impurity electrons leave their parent atoms and drift within the material. See fig. 7.6(1).

Some electrons in the n-type material eventually drift across the junction between the two extrinsic semiconductors and continue their random movement at the other side of the junction. Some then return but others complete the incomplete bonds in the p-type material, i.e. they 'fill the holes'. Having done so these electrons are bound within the material, the p-type material (which was neutral) has gained electrons and the n-type material (which also was neutral) has lost electrons. See fig. 7.6(2).

The p-type material now has an overall negative charge and the n-type material has an overall positive charge. There is therefore a difference in electrical potential across the junction. It is called the *junction p.d.* and has a value of approximately 0.2 V for germanium and 0.6 V for silicon.

Once the junction p.d. is set up, the p-type material, being now negatively charged, repels further random movement of electrons across the junction. For this reason the junction p.d. is sometimes called the *barrier potential*. The area within both materials close to the junction is called the *depletion zone* since it is depleted of carriers, the holes close to the junction in the p-type material having been filled by electrons which moved from the region close to the other side of the junction in the n-type material.

Diode action is now possible for the following reasons. If an external p.d. is applied to the material such that the p-type material is made negative and the n-type material is made positive, the junction p.d. set up by random movement is reinforced and the barrier effect strengthened. Electron flow from n-type to p-type is thus stopped.

A small number of free electrons exist in the p-type from bonds which break up and reform due to heat energy absorption (minority carriers) and some of these are attracted to the n-type which is now electrically positive. A small current therefore flows. If, however, the external p.d. is applied, positive to p-type, negative to n-type, provided it is sufficiently high to overcome the junction p.d. (greater than 0.2 V or 0.6 V for germanium or silicon respectively), the junction p.d. is neutralised and the n-type material electrons (majority carriers), which are loosely bound to the pentavalent impurity atoms, break free and a much larger current flows.

When the external p.d. is applied so that minimum current flows (negative to p-type, positive to n-type), the diode is said to be *reverse biased*. When the external p.d. is applied so that maximum current flows (positive to p-type, negative to n-type), the diode is said to be *forward biased* (see fig. 7.6). Note that the p-type material is called the *anode* and the n-type material is called the *cathode* of the diode.

A typical current–voltage characteristic of a semiconductor diode is shown in fig. 7.7a. When the cathode of the semiconductor diode (the n-type material) is made *positive* with respect to the anode (the p-type material) a small current flows due to the minority carriers in the p-material, free electrons which are there because of atomic bonds breaking and reforming. This breaking-up and recombination is due to absorption of heat energy from the diode surrounding so the small reverse current is temperature-dependent as shown.

If the reverse p.d. is increased, a point is reached where the electric field across the junction is so strong that the atomic structure of the crystal breaks down and a large current flows. This is called the *Zener effect*. The breakdown may be increased further by the moving electrons being accelerated and acquiring much more energy, these electrons then colliding with others within the crystal structure and releasing them.

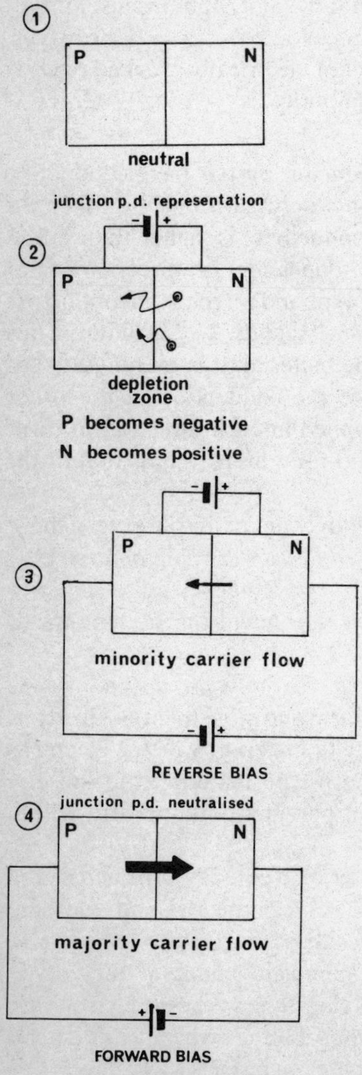

Figure 7.6

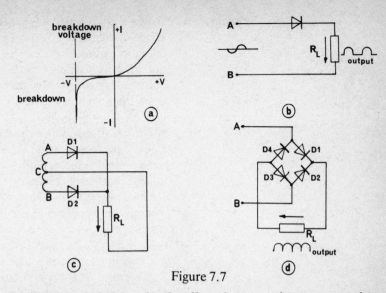

Figure 7.7

This is called the *avalanche effect*, since one electron may release two more which each release two more and so on. Unless the diode is specially made to withstand this it may be permanently damaged if breakdown is allowed to occur. A p.d. equal to the breakdown voltage, called the *peak inverse voltage*, should not therefore be applied to the diode in the reverse bias connection. Certain diodes are specially made to operate in this region for circuits requiring a variable current–constant voltage characteristic, for, as shown in fig. 7.7a, the breakdown occurs at a relatively constant voltage. These special semiconductor diodes are called *zener diodes* or *avalanche diodes*.

Comparing silicon diodes with germanium diodes, the junction p.d. (barrier potential) which must be overcome for forward conduction to occur is larger for silicon than for germanium diodes. Examination of actual diode characteristics also shows that the current which flows under reverse bias conditions is smaller in silicon diodes because of there being fewer free electrons in the n-type material. As was said earlier, these free electrons are produced by absorption of heat energy, the temperature of the diode surroundings having a direct effect.

In general, silicon components tend to be less affected by external temperature than germanium components and the smaller reverse current in a silicon diode for the same operating conditions is an example of this lesser dependence on temperature. It should be noted that temperature also affects the forward current of semiconductor diodes, again silicon diodes being affected less per degree of temperature change than germanium diodes.

Silicon diodes are able to operate at higher temperatures than germanium diodes, between two or three times as high, in fact, and accordingly tend to be used more in higher temperature surroundings and also at higher power levels. Both kinds of diode are available for use at low voltage and current (signal diodes) and at high voltage and current (power diodes).

Summary   A diode is an active component having two electrodes, called the anode and cathode. When a p.d. is applied across these electrodes the current flow depends upon the size and polarity of the anode potential with respect to the cathode potential. When the anode potential is more positive than the cathode potential a larger current flows than when the polarity of the anode is reversed. When the larger current flows, the diode is said to be forward biased. When the polarity is reversed, the anode potential being less positive than the cathode potential, the diode is said to be reverse biased.

The semiconductor diode works on the principle of adding minute quantities of special impurities to pure (intrinsic) crystals of either silicon or germanium. This process is called doping and changes the conductive properties of the crystal. There are two types of doped (extrinsic) crystal, p-type with electron-vacancies or holes, and n-type with impurity electrons being only loosely bound to their parent impurity atoms. When p-type and n-type materials are alongside one another, a p.d. is set up across the junction by the movement of charge carriers, due to their absorbing energy from their surroundings. If an external p.d. is connected so as to reinforce the junction p.d. only a small current flows (reverse bias). If an external p.d. is connected so as to oppose the junction p.d. a larger current flows (forward bias). The two-layer crystal thus behaves as a diode.

The semiconductor diode conducts slightly in the reverse direction, the current being very small provided the reverse voltage does not exceed a critical value called the breakdown voltage. At this voltage zener or avalanche breakdown occurs and the reverse current rises rapidly, the reverse voltage remaining approximately constant. This constant voltage–variable current effect in the reverse direction of a semiconductor diode may be used in voltage regulator circuits. Of the two types of semiconductor diode, silicon diodes are less temperature sensitive than germanium and tend to be used in higher power circuits.

General objective   *The expected learning outcome is that the student relates the concepts of a.c. theory to an elementary treatment of half- and full-wave rectification.*

Specific objectives   *The expected learning outcome is that the student:*
7.1 Defines the elementary principles of half- and full-wave rectification as the conversion from alternating to unidirectional voltage, by the application of simple switching.

Rectification   Rectification is the process of converting an alternating current (one which flows in both directions) into a direct current (one which flows only in one direction).

Figs 7.7b and 7.7c show simple rectification circuits using diodes. In fig. 7.7b a single diode is connected in series with a resistive load R_L, the alternating supply voltage being applied across the series circuit. When point A is positive with respect to B the diode conducts and a relatively large current flows in the load; when point A is negative with respect to B the diode is reverse biased and only a very small current flows. The voltage waveform across R_L is as shown and as can be seen only one half of the input waveform appears at the output virtually unchanged, the other half being reduced almost to a point where it can be ignored. The circuit is called a half-wave rectifier circuit.

The circuits in figs 7.7c and 7.7d are called full-wave rectifier circuits and allow both halves of the input voltage waveform to appear at the output such that they act in the same direction across the load. The circuit of fig. 7.7c uses only two diodes but requires a centre-tapped transformer. When point A is positive with respect to point B, conventional current flows from A through D1, through the load from top to bottom in the figure, and then to the centre tap C. Diode D2 is reverse biased during this period. When point B is more positive than point A, diode D2 conducts and again current flows in the load from top to bottom as shown. Diode D1 is reverse biased on this occasion. The output waveform consists of a series of half sinusoid pulses as shown in fig. 7.7d.

The circuit of fig. 7.7d shows a full-wave rectifier requiring four diodes but not requiring a centre-tapped transformer. Here, when A is positive with respect to B, current flow is via D1, R_L and D3 and in the other half-cycle, when B is positive with respect to A, via D2, R_L and D4. The same shaped waveform is obtained as in the previous circuit since current flow on the load is always in the same direction. This circuit is sometimes called a *bridge rectifier* circuit and four diodes permanently mounted on the same shaft for use in this kind of circuit are collectively referred to as a bridge rectifier.

Specific objectives *The expected learning outcome is that the student:*
10.1 *Sketches the physical construction of typical n-p-n transistors.*
10.2 *Identifies the emitter, base and collector for p-n-p and n-p-n transistors.*
10.3 *Identifies a forward biased base–emitter junction.*
10.4 *Identifies a reverse biased base–collector junction.*
10.5 *Draws a sketch for basic transistor action to show forward biased base–emitter junction and reverse biased base–collector junction.*
10.6 *Describes the transistor action for a p-n-p arrangement.*
10.7 *Describes the transistor action for n-p-n arrangement.*
10.8 *Describes the leakage current effects for 10.7.*
10.9 *Identifies the p-n-p and n-p-n transistor symbols.*

Transistors As was discussed above, the diode, which is a two-layer semiconductor device, conducts better in one direction than in the other. To control current flow in a diode the potential of one of the connections is changed with respect to the other and control is exercised at a point within the same circuit of which the diode is part.

The transistor has three connections and the current flow between two of them can be controlled by the third. Current flow in a circuit can thus be controlled by a separate circuit. Further, the controlled current may be many times larger than the control current and since there is a definite relationship between the currents and a change in one is reflected into the other, the transistor is able to amplify (make larger) electronic signals. Diodes are not able to do this; together with other components such as resistors, capacitors, etc. diodes may be used to change signal waveforms but the signal is usually attenuated (reduced) in the process.

There are two kinds of transistor, the *bipolar* transistor and the *field-effect* transistor (FET). Both kinds do the same sort of job but they work in a different way. Both kinds are available as separate (discrete) components or may form part of an integrated circuit in which all the components (diodes, transistors, resistors and capacitors) are formed within a single piece or *chip* of silicon. In this section we shall be concerned with the bipolar transistor; the field-effect transistor is described in a later unit.

As we have seen, conduction in a semiconductor may occur due to the movement of electrons or to the movement of holes. Hole conduction too depends upon electron movement, the electrons in this case being valency electrons rather than the more loosely bound outer electrons. The bipolar transistor uses both kinds of conduction, hence the name bipolar. The FET uses one kind only, which is why this transistor is sometimes referred to as a unipolar transistor.

A bipolar transistor consists of three layers of doped semiconductor called the *emitter*, *base* and *collector*. The base is a very thin layer, which is essential for transistor operation, and lies between the other layers. The base may be either p-type or n-type, the collector and emitter being of the opposite type to the base so that we may have a p-n-p transistor or n-p-n transistor. Both types do the same job, the difference is in the polarity of the connections from the circuit supply voltage.

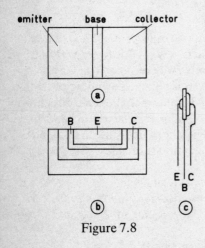

Figure 7.8

To explain transistor action the diagram of fig. 7.8a will be used. This arrangement is similar to the *planar* form of construction shown in fig. 7.8b but may also represent the *alloy-junction* form of construction shown in fig. 7.8c. The planar transistor, which is now the most common type, may be a discrete component or may form part of an integrated circuit. The alloy-junction method of construction is not used in integrated circuits.

For both n-p-n and p-n-p types of transistor the method of operation is the same. For conduction to take place between emitter and collector, the emitter–base junction is *forward* biased (positive to p-type, negative to n-type) and the base–collector junction is *reverse* biased. Operation of an n-p-n type of transistor will be described first (see fig. 7.9).

An n-p-n transistor is shown diagrammatically in fig. 7.9a. The symbol used in circuit diagrams is shown in fig. 7.9b. Biasing is obtained here by using two batteries (or two parts of the same battery) and this method may be used in practice or alternatively, as is shown later in the chapter, a single battery or voltage source and a *bias resistor* may be used.

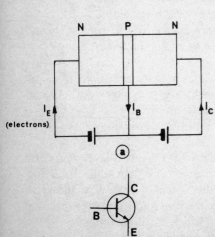

Figure 7.9

In the figure, the emitter–base junction is forward biased by making the base *positive* with respect to the emitter. The base–collector junction is reverse biased by making the collector more positive than the base. Under the influence of the forward bias, relatively large numbers of electrons are injected into the emitter and move across the junction into the base layer. If this were a diode these electrons would then leave the p-type material and return to the battery, but because the collector region is at a positive potential with respect to the base and because the base layer is extremely thin, the majority of the electrons move across the base layer and into the collector region, only a small proportion leaving the base to return to the emitter–base battery.

The larger part of the emitter current thus moves into the collector and these electrons return instead to the positive side of the base–collector battery. If the base–collector junction were a diode only a small current would flow because of the reverse bias. A large current flows through the junction in a transistor because of the relatively large numbers of electrons injected into the thin base region from the emitter.

There are then *three* currents, the emitter current, the base current and the collector current, the emitter current being the *sum* or the other two. The collector current is almost equal to the emitter current but is slightly smaller, the difference between the two being the base current. Both collector and emitter current are many times larger than the base current.

Fig. 7.10 shows a p-n-p transistor and its circuit symbol. Here again the emitter–base junction is *forward* biased and the base–collector junction *reverse* biased, but because the type of semiconductor is opposite in each region to that in the n-p-n transistor, the bias batteries are connected the other way round. The base is *negative* with respect to the emitter, the collector even more negative and thus negative with respect to the base.

In the p-n-p transistor the main type of conduction is hole conduction, large numbers of holes moving from emitter to base and then to collector. A small proportion of holes is filled in the base region by electrons moving into the base and again, as with the n-p-n transistor, a small number of electrons are moving in the base lead (this time *into* the base not out of it as with the n-p-n transistor) and a much larger number of electrons are moving into the collector to fill the large number of holes which accumulate there. (The idea of 'hole' conduction cannot be applied to circuit leads which are usually made of copper or some similar material; holes exist only in a semiconductor material because of its particular kind of crystalline structure.)

So again with the p-n-p transistor, there are three currents as before, the emitter current being the sum of the other two. The difference is only that within the p-n-p transistor, conduction is mainly of the 'hole-conduction' type, i.e. due to valency electron movement, whereas in the n-p-n transistor the conduction in the transistor is the more 'normal' conduction-electron type.

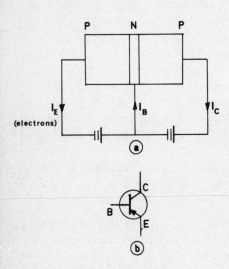

Figure 7.10

For both types of transistor, $I_E = I_B + I_C$

where I_E, I_B and I_C represent the emitter, base and collector currents respectively.

This equation is true at all times. When the currents are continuously varying, as they are in amplifier circuits, for example, the equation is still valid at any instant.

If the emitter–base bias is altered, the emitter and base currents change causing a corresponding change in the collector current. This change is much larger than the change in base current and the transistor can be used for controlling a circuit current by means of a much smaller current. It can therefore be used for remote switching (controlling one circuit by a separate circuit at lower voltage and current levels) or for amplification. Amplification of an electronic signal means making the signal larger while retaining the same waveform. It is possible because over a particular range of frequencies the collector current change follows the base current change quite faithfully.

A graph of collector current plotted against base current over the frequency range of a typical transistor is shown in fig. 7.11. As shown it is a straight line which tells us two things; over the range shown:

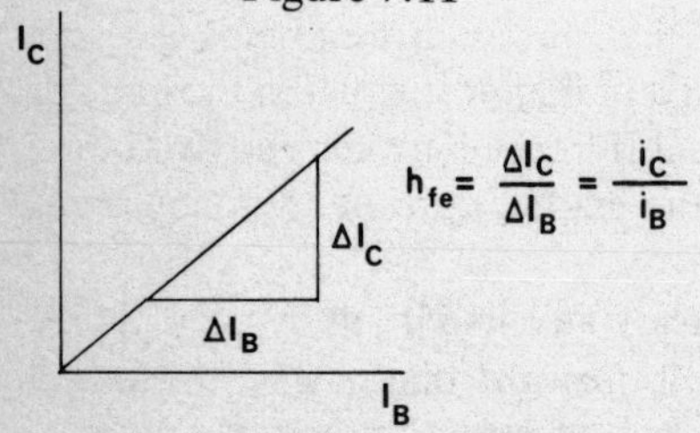

(1) the ratio I_C/I_B is constant, i.e. at any point on the graph, dividing the value of I_C by the corresponding value of I_B gives the same value.
(2) the gradient is constant so that, when there is a change in base current and a corresponding change in collector current, dividing the *change* in collector current by the *change* in base current gives the same result whatever the value of I_C and I_B about which the change takes place. See fig. 7.12.

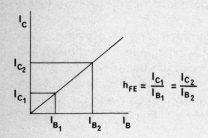

Figure 7.11

Figure 7.12

The ratio I_C/I_B is the d.c. current gain of the transistor and is given the symbol h_{FE}.

The ratio $\dfrac{\text{change in } I_C}{\text{change in } I_B}$, written symbolically as $\dfrac{\Delta I_C}{\Delta I_B}$,

is the a.c. or small signal current gain of the transistor and is given the symbol h_{fe}. The symbols h_{FE} and h_{fe} are *parameters* of the transistor, i.e. measured characteristics, and differ from transistor to transistor. There are a number of kinds of transistor parameter, these two are called 'h parameters'.

For the transistor giving the I_C/I_B graph shown, h_{FE} and h_{fe} have the same value. This is usually true over a limited range of currents and of frequency. At higher frequencies the d.c. current gain and a.c. current gain tend to be different in value.

Leakage current in transistors

Transistors have many advantages over earlier electronic devices used for switching and amplification. They are very small, robust and extremely reliable. They require only small voltages and currents and do not use heaters as did electronic valves. However, they do have one disadvantage of particular importance in the bipolar transistor. This is that to some degree they may be affected adversely by temperature, both of the transistor itself and of the air or other medium surrounding the transistor.

To understand why this is so we must take a closer look at what is happening in a p-n junction when it is reverse biased. As has been stated both p-type and n-type materials are in themselves electrically neutral but each material is able to conduct (in either direction) because of either a large number of holes (p-type) or a large number of electrons (n-type) which have been made available by the addition of impurities to the pure semiconductor.

However, both electrons and holes are present in both types of material all the time, due to the crystal hole–electron bonds breaking and reforming. The breaking occurs because electrons absorb heat energy and leave their bonds. Usually they do not move far and recombination takes place so that a state of equilibrium is reached.

When a p-n junction is formed some electrons move randomly from n-type to p-type and a junction p.d. is set up as was explained earlier. Forward biasing the junction neutralises this p.d. and a large electron flow from n-type to p-type occurs. When reverse bias is applied the junction p.d. is reinforced and this current flow stops. However, because there *are* a minority of electrons available in the p-type due to thermal breakdown of the atomic bonds, these electrons move from p-type or n-type under the influence of the reverse bias field set up by the applied voltage, and a small reverse current flows. Thus reverse current is temperature dependent.

In the bipolar transistor the base–collector junction is reverse biased and a small temperature dependent current flows into the base even when the emitter lead is not connected. When it is connected and transistor action occurs a larger base current flows, part of which is this temperature-dependent current. The temperature-dependent current is called the *leakage current* of the transistor. The collector current is many times larger than the base current, i.e. amplification takes place, so that a larger part of the collector current is temperature dependent – the leakage current is effectively amplified.

The transistor temperature depends not only on its surroundings but also on the power dissipated at the collector by the collector current. If the transistor temperature rises the leakage current rises and the collector current rises even more, causing the transistor temperature to rise even further. The process is cumulative and if unchecked the collector current will rise until the heat generated destroys the transistor. The process is called *thermal runaway*.

Special arrangements of bias circuits are used in practice to detect any untoward increase in base current due to anything other than signal change and thermal runaway is avoided. However, these circuits only have a limited temperature range themselves and it is advisable to handle electronic equipment taking into account manufacturer's instructions concerning temperature and environment. Electronic circuits for laboratory instruments or domestic or industrial use are not built to withstand the same hazardous conditions as those built for use for military and aero-space use!

142 Electrical and Electronic Principles 2

Specific objectives

The expected learning outcome is that the student:

10.11 Explains the electron current flow paths using the p-n-p and n-p-n transistor symbols.

10.12 Describes how transistors can be connected in the following configurations:
 (a) common base,
 (b) common emitter,
 (c) common collector (emitter follower)

Using the transistor

Almost all electronic circuits are used for the control and processing of data or information and as such they have *input* connections and *output* connections.

The transistor has three connections so that if two are used for input and two are used for output, one must be common to both input and output. There are *three* kinds of circuit – common emitter, common base and common collector. See fig. 7.13.

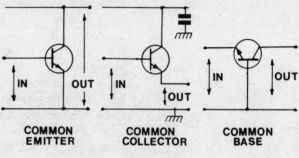

Figure 7.13

The lower supply line is often referred to as 'ground' (although it may or may not be connected to earth) and hence these circuits are sometimes referred to as grounded emitter, grounded base and grounded collector respectively.

The common emitter circuit

Fig. 7.14 shows a simple common emitter circuit which may be used for switching or, if the operating conditions are chosen correctly, for signal amplification.

The transistor shown is n-p-n (but it could equally be p-n-p if the polarity of connections is reversed). Electron flow is from the emitter to the collector and then via the load to the positive supply line with a smaller flow from the base via resistor R_B to the positive supply line. Resistor R_B is the *base bias resistor* and the p.d. across it is the supply voltage LESS the emitter–base forward bias V_{BE}. The load may be a resistor or any other kind of component or device in which the current is to be controlled.

To use the circuit as a remote switch the base bias resistor connection may be opened, in which case the base current and the resultant collector current are both reduced to zero. When the bias resistor is connected the base current (of the order of μA or mA) flows and the collector current (mA or A) flows and the load is activated. Thus a small current is used to control a larger one.

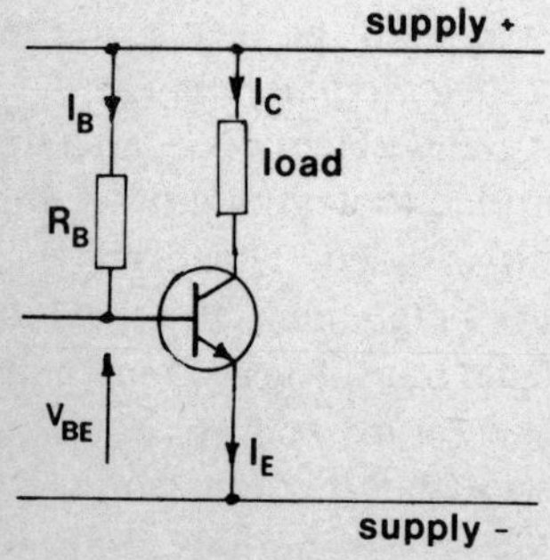

Figure 7.14

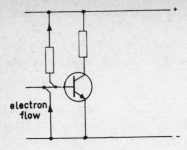

Figure 7.15

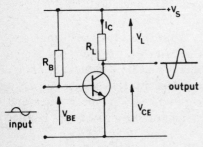

Figure 7.16

Alternatively, the base connection may be short-circuited to ground (the emitter in this case). The electron current in the base bias resistor now flows from the negative side of the supply to the positive side, avoiding the emitter–base path; there is no base current and no collector or load current (Fig. 7.15).

Low current/voltage switching is commonly used on control panels to avoid danger to the operator, the circuits eventually being controlled working at many hundreds or even thousands of volts (motor control, control of the National Grid distribution system of electricity, which works at voltages in excess of 250 000 V).

Other examples of switching occur in transistor logic circuits which form the basis of computer and microprocessor circuits.

This form of circuit may also be used as an amplifier. Suppose the load is a resistor R_L, the p.d. across it being V_L.

From fig. 7.16, $V_S = V_{CE} + V_L$ and $V_L = I_C R_L$

so that $V_S = V_{CE} + I_C R_L$

Now, if a signal is applied between base and emitter causing the potential at the base with respect to the emitter to rise and fall, the base current value will rise and fall as will the collector current. The p.d. V_L, which is equal to $I_C R_L$, will also rise and fall.

Now V_S is constant, so that since $V_S = V_{CE} + V_L$ and V_L is rising and falling, V_{CE} must be falling and rising.

If operating conditions are correctly chosen, the change in V_{CE} will follow the change in V_{BE} faithfully but will be a larger or amplified version of it and the transistor is acting as a voltage amplifier. Note that the change in V_{CE} is in the opposite sense to the change in V_{BE}, i.e. there is a 180° phase shift between output and input.

The common base circuit

In this circuit the input is connected between emitter and base, the output between collector and base. There is no current gain, since the collector current is less than the emitter current, but the circuit may nevertheless be used to amplify voltage.

Two important characteristics of an amplifier are the impedance which it presents to an input signal, called the *input impedance* and the impedance it presents at its output, called the *output impedance*. A common base amplifier has a very low input impedance and high output impedance.

The common collector circuit

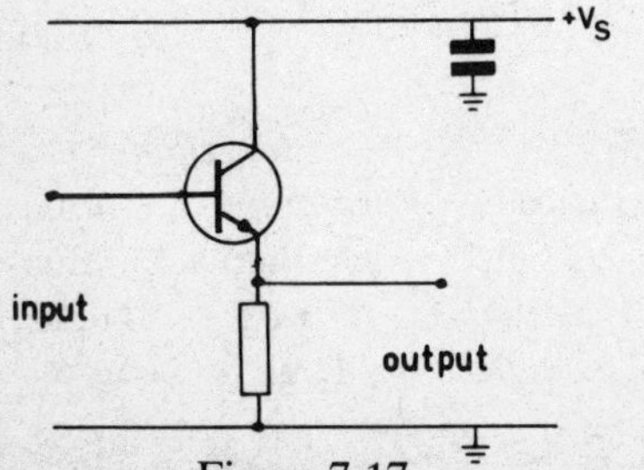

A common collector circuit or *emitter follower* is shown in fig. 7.17. Here the collector is connected to ground as far as the signal is concerned – usually by a capacitor offering a low reactance path to signal – and the input, which is between base and ground, is effectively between base and collector. Similarly, the output, which is between emitter and ground, is effectively between emitter and collector.

As can be seen, the emitter–ground voltage is less than the base–ground voltage by an amount equal to the base–emitter voltage so the circuit is *not* a voltage amplifier. It is however a current amplifier,

Figure 7.17

since the output (emitter) current is much larger than the input (base) current.

Also, the circuit has a high input impedance and very low output impedance which are extremely useful in some applications. These are discussed in subsequent units.

Summary

A transistor is a three-electrode semiconductor component capable of amplifying electronic signals. It may also be used for switching circuits in which a small current is used to control a larger current.

There are two types of transistor, the bipolar transistor and the field-effect transistor (FET). The bipolar transistor uses both electron conduction and hole conduction in its operation. The FET uses one or the other type of conduction but not both.

The bipolar transistor consists of three layers of semiconductor material, the centre layer, called the base, being of the opposite type of semiconductor to that of the other two layers. The outer layers are called the emitter and the collector respectively. There are thus two types of bipolar transistor, the p-n-p, with p-type emitter and collector and n-type base and the n-p-n transistor with n-type emitter and collector and p-type base.

For conduction the emitter–base junction is forward biased, the base–collector junction is reverse biased. When biased in this manner there are three currents associated with the transistor, the emitter current, the base current and the collector current. The emitter current is equal to the sum of the other two currents.

The collector current is many times larger than the base current, the ratio between the d.c. values of these currents (collector current/base current) being called the d.c. current gain, symbol h_{FE}. If the base current changes the collector current also changes, the ratio between the changes (collector current change/base current change) being called the a.c. current gain, symbol h_{fe}.

The bipolar transistor may be used in any of three modes determined by where the input signal is connected to and output signal taken from:

Input	Output	Name of mode
Base–emitter	Collector–emitter	Common emitter
Emitter–base	Collector–base	Common base
Base–collector	Emitter–collector	Common collector (Emitter-follower)

Each mode of connection has its own particular characteristics of voltage and current amplification and of input and output impedance.

| Mode | Gain | | Impedance | |
	Voltage	Current	Input	Output
Common emitter	High	High	Medium	Medium
Common base	High	Less than 1	Low	High
Common collector	Less than 1	High	High	Low

Bipolar transistors are temperature sensitive and special biasing circuits are used to protect them from effects due to rising temperature. Without such protection a bipolar transistor may destroy itself due to a cumulative rising temperature—rising current process called thermal runaway.

EXERCISE 7

1. State how the external p.d. is connected to a diode:
(a) for forward bias; (b) for reverse bias.

2. Is silicon affected more or less than germanium by the temperature of the surroundings?

3. Is the junction p.d. of a silicon diode higher or lower than that of a germanium diode?

4. Which kind of semiconductor diode is more likely to be used in high-power circuits.

5. Define the term 'depletion zone'.

6. Why is the junction p.d. of a semiconductor diode called the barrier potential?

7. Extrinsic semiconductor containing arsenic;
A. is called p-type semiconductor;
B. has a smaller conductivity than intrinsic semiconductor;
C. conducts equally well in all directions;
D. has electron-absences or 'holes' in its atomic structure.

8. The junction p.d. of a silicon diode lies between
A. 0 V and 0.15 V;   C. 0.3 V and 0.45 V;
B. 0.15 V and 0.3 V;   D. 0.45 V and 0.7 V.

9. Using diagrams if necessary, describe how the conducting properties of a pure semiconductor may be altered by the addition of impurities.

10. Sketch and explain the current/voltage characteristic of a semiconductor diode. Include in your answer the meaning of the terms 'breakdown' and 'peak inverse voltage'.

11. Name the three electrodes of a bipolar transistor.

12. Name two types of bipolar transistor.

13. The correct biasing for conduction in a bipolar transistor for the emitter–base and base–collector junctions, is respectively:
A. forward, forward; B. forward, reverse; C. reverse, forward; D. reverse, reverse.

14. For a bipolar transistor:
A. h_{fe} represents d.c. current gain;
B. h_{FE} represents a.c. current gain;
C. h_{fe} and h_{FE} are always equal;
D. h_{fe} and h_{FE} represent the a.c. and d.c. current gain respectively.

SELF-ASSESSMENT EXERCISE 7

	Possible marks
1. State the names of the connections to a diode.	(3)
2. Name the majority carrier in n-type semiconductor.	(3)
3. State two properties of a semiconductor.	(3)
4. State the valency of the impurity in a p-type semiconductor.	(3)

5. What is meant by the term 'intrinsic semiconductor'? (3)
6. Select the correct statement:
   A. P-type semiconductor contains a pentavalent impurity.
   B. A forward-biased p-n diode has the p-region negative with respect to the n-region.
   C. Reverse bias adds to the barrier potential in a p-n diode.
   D. Forward bias creates a depletion zone in a p-n diode. (5)
7. Draw a diagram showing how a p-n diode is:
   (a) forward biased; (b) reverse biased. (5)
8. Draw the symbols of:
   (a) p-n diode; (b) p-n-p bipolar transistor; (c) n-p-n bipolar transistor (5)
9. Explain the basic theory of operation of a semiconductor diode. Include in the explanation details of the materials, construction and connection in circuit. (14)
10. (a) Draw the symbols for the two types of bipolar transistor;
    (b) Describe the operation of ONE of these types of transistor; (14)
11. Describe and compare the three modes of operation of a bipolar transistor. (14)
12. (a) Explain what is meant by the 'a.c. current gain' of a transistor;
    (b) Explain the difference between the transistor parameters h_{fe} and h_{FE};
    (c) Describe briefly how a transistor can be used as a switch. (14)
13. A bipolar transistor is sometimes described as 'two junction diodes back to back'. Explain why this description is used and why two actual diodes connected in such a manner do not actually function as a transistor. (14)

Answers

SELF-ASSESSMENT EXERCISE 7

Marks

1. Anode, cathode. ($1\frac{1}{2}$ each)

2. Electrons. (3)

3. A semiconductor has a resistivity lying between that of a conductor and an insulator, it also has a negative temperature coefficient. ($1\frac{1}{2}$ each)

4. Three (a tri-valent impurity) (3)

5. A pure semiconductor (3)

6. C (5)

7. Diagram showing:
Forward bias: positive to p-type; negative to n-type.
Reverse bias: negative to p-type; positive to n-type ($2\frac{1}{2}$ each)

8. Symbols (see text). ((a) 1 (b), (c) 2 each)

9. Explanation should include the following points:
(i) Description of crystalline tetravalent structure of a semiconductor; (2)
(ii) Addition of trivalent and pentavalent impurities; (3)
(iii) What happens in an unbiased p-n junction at room temperature (random movement of carriers, formation of barrier potential and depletion zone). (3)
(iv) What happens when the junction is forward biased (including *how* a junction is forward biased, removal of barrier potential, majority carrier flow etc.) (3)
(v) What happens when the junction is reverse biased (strengthened carrier potential, minority carrier flow etc.) (3)

10. (a) Symbols: see text (1 each)
    (b) Description must include:
        (i) Construction (emitter, thin base layer, collector); (2)
        (ii) Forward bias of emitter–base; reverse bias of base–collector; (2)
        (iii) Majority carrier flow from emitter to base, most of which continues to collector because of thin base layer; (3)
        (iv) Base current very small, collector current larger; (2)
        (v) Change in base–emitter voltage changes base and collector currents; ratio of changes I_C/I_B is current amplification. (3)

11. The three modes are common emitter, common base and common collector. The description for *each* mode must include details of where the input is connected, from where the output is taken, details of input and output impedance:
    Mode name (2)
    Input; output; input impedance; output impedance (3 each)

12. (a) A.C. current gain is the ratio of the *change* in output current to the corresponding *change* in input current (2)

    (b) $h_{fe} = \dfrac{\Delta I_C}{\Delta I_B}$; $h_{FE} = \dfrac{I_C}{I_B}$ where I_C is collector current (d.c. value), I_B is base current (d.c. value) and 'Δ' means 'change in'. (3 each)

    The choice of upper and lower case letters (capitals and 'small' letters) MUST be correct; otherwise the answer is wrong.

    (c) Description must include the fact that:
        (i) The base–emitter voltage controls the collector current making it smaller, larger or zero; (2)
        (ii) When used as a switch the base–emitter voltage is used to switch the collector current from zero to some other value; (2)
        (iii) Low voltage/current can be used to switch high voltage/current circuits. (2)

13. Explanation must include:
    (i) A junction diode consists of two layers, one p-type, one n-type, called anode and cathode respectively; (2)
    (ii) A bipolar transistor consists of three layers, n-p-n or p-n-p, called emitter, base and collector; (3)
    (iii) The emitter–base junction in a p-n-p transistor can be likened to the anode–cathode of a diode (or in an n-p-n transistor to the cathode–anode of a diode); (3)
    (iv) The base–collector junction of a p-n-p transistor can be likened to the cathode–anode of a diode (or of an n-p-n transistor to the anode–cathode); (3)
    (v) Bipolar transistor action depends on there being three adjacent layers, the middle one (base) being very thin. Two actual diodes would not provide this construction. (3)

8 Measuring instruments and measurements

Topic area: I

General objective The expected learning outcome is that the student understands the operation, uses and limitations of measuring instruments.

Specific objectives The expected learning outcome is that the student:
11.1 Describes, with the aid of given diagrams, the principles of operation of moving-coil instruments.
11.2 Uses ammeters and voltmeters correctly in d.c. circuit measurements.
11.3 Explains the need for shunts and multipliers to extend the range of a basic electrical indicating instrument.

There are two kinds of electrical indicating and measuring instruments, electromechanical and electronic. In the first kind there are moving parts and the quantity being measured causes a deflection of a pointer along a calibrated scale or, sometimes, the deflection of a calibrated scale past a fixed pointer. The deflection is proportional to the measured quantity and is produced using electrical or electromagnetic means.

Electronic instruments use electronic circuitry to measure the quantity and the measured value is displayed using electron tubes or electronic display devices. There are also hybrid (mixed) instruments in which electronic circuitry measures the quantity but the value is displayed using moving pointers or moving scales.

Electromechanical instruments There are three essential parts to any electromechanical instrument. These parts produce respectively:
(a) deflection; (b) control; (c) damping.

(a) *Deflection*. Part of the instrument allows the quantity being measured to set up a deflecting force so that the pointer (or scale) moves. The deflecting force must in some way be proportional to the measured quantity.

Deflection is normally produced by the force caused by interaction between magnetic fields, at least one of these being set up by the quantity being measured. There are three types of such instrument in common use, the moving-iron, moving-coil and electrodynamic or dynamometer type.

The moving-iron instrument In the moving-iron instrument the pointer is attached to a piece of soft iron which is able to move freely and which is situated either inside or close to a coil through which a current is passed. The current is caused by the measured quantity.

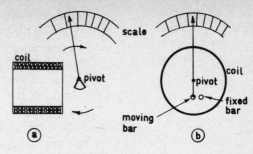

Figure 8.1

Fig. 8.1a illustrates the attraction type of moving-iron instrument in which the magnetic field set up by the coil attracts the soft iron piece and moves the pointer accordingly. In the repulsion type, shown in part (b) of fig. 8.1, there are two pieces of soft iron, of which one is fixed and one is able to move. Here both pieces of iron are magnetised with the same polarity by the coil field and a repulsive force is set up causing the pieces to move apart. In both types, the soft iron pieces lose their magnetism when the coil current falls to zero and the pointer returns to the rest position.

In both types of instrument the deflection is proportional to the square of the coil current so that equal increments of current do not produce equal increments of deflection and the scale is *non-linear*, i.e. it is one in which the calibration marks are closer together at one end than at the other. Specially-shaped pole pieces may be used to improve scale linearity but the scale remains essentially non-linear.

Moving-iron instruments may be used as ammeters or voltmeters and, since the deflection is in the same direction regardless of the direction of flow of the coil current, they may be used to measure either a.c. or d.c. They are used mainly as a.c. instruments and their chief advantages are ruggedness, simplicity of construction and they are relatively cheap to manufacture. The range of current or voltage being measured may be extended beyond that which produces the maximum deflection of the pointer called the *full scale deflection* or f.s.d. value, using instrument transformers.

The permanent-magnet, moving-coil (p.m.m.c.) instrument

This instrument, sometimes called a d'Arsonval instrument, uses a coil which is free to move within the magnetic field of a permanent magnet (see fig. 8.2).

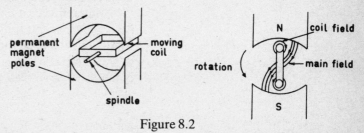

Figure 8.2

A current produced by the quantity being measured flows in the coil and the magnetic field set up by the coil current reacts with the main permanent-magnet field, as described in chapter 4, to produce a deflecting force which is directly proportional to the coil current. The coil being free to move, does so and begins to rotate about the coil spindle. The indicator, pointer or scale, is fixed to the spindle and moves with it.

In the p.m.m.c. instrument the torque (turning force) on the coil caused by the coil current, is directly proportional to it. The calibration marks on the scale indicating the value of the measured quantity are thus equally displaced for equal increments of coil current and the scale is linear, provided, of course, that the coil current is directly proportional to the quantity being measured.

The direction of deflection depends upon the direction of flow of the coil current since this, in turn, determines the direction of the coil magnetic field and the permanent-magnet magnetic field is fixed. Alternating current must therefore be rectified before passing it through the instrument. The range of the movement may be extended using series or parallel resistors depending upon the use for which the instrument is intended; this is described in more detail below.

P.M.M.C. instruments are used widely for the measurement of current, voltage and resistance and, with the aid of suitable transducers (to obtain an electrical signal), to measure a large number of physical quantities in industry and in the laboratory. Their chief advantages include the linear scale, ease of extension of ranges and the low power consumed by the instrument.

If a second coil is used instead of the permanent magnet to produce a magnetic field to react with the field produced by the measured quantity the instrument is called an electrodynamic instrument or dynamometer. This is further described later.

(b) *Control*. As the deflection takes place a controlling force is set up dependent on the deflecting force so that when the two are equal the movement of the pointer or scale stops, the final deflection then being proportional to the measured quantity.

There are two main types of control mechanism in use, spring control and taut band control. In the first method the control force is established by means of a coiled spring which when tightened or loosened from its normal rest position attempts to return to it, thus setting up an opposing or control force on the moving pointer. Usually, two springs are used one at either end of a spindle on which the pointer or scale is mounted and these springs are *contra-wound*, i.e. wound in opposite directions, so that the coil movement tends to wind, or tighten, one spring and unwind, or loosen, the other.

This method of using the springs helps to compensate for temperature effects which may affect a single spring and change its characteristics. Temperature effects on one spring will be opposite to those on the other, if the springs are identical and contra-wound, and the overall temperature effect is neutralised. When spring control is used with a p.m.m.c. instrument the springs are also used as leads for the coil current to be taken into and out from the coil.

Taut band control uses a ribbon of metal which suspends the

moving part of the instrument. When deflection occurs the ribbon or band is twisted and tends to return to its rest position in a similar way to a spring. The restoring force acting on the band also acts as a controlling force on the movement and again when the deflecting force and controlling force are equal the pointer stops moving.

(c) *Damping*. Without some form of damping the moving part will move past the final rest position so that the control force will be greater than the deflecting force and the moving part then reverses its movement. It then again moves past the rest position, in the opposite direction, so that the deflecting force is greater than the control force and the direction of movement is again reversed. The pointer or scale will oscillate about the final position.

The natural damping of the system (due to friction or air resistance or some other factor) is not usually enough to prevent oscillation and some additional method of damping is required. Fig. 8.3 shows the movement of an instrument pointer or scale under four different conditions. An underdamped instrument oscillates, an overdamped instrument takes too long to settle into the final position. Critical damping, which at first sight may appear ideal, allows the indicator to move fairly quickly to the rest position and stop. This damping is not ideal, however, because exactly the same result would occur if for some reason the indicator stuck at the same point along its path. Ideally, the indicator should move slightly *past* the rest position and then return to it without undue oscillation.

Some methods of damping are shown in fig. 8.4.

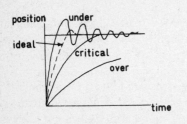

Figure 8.3

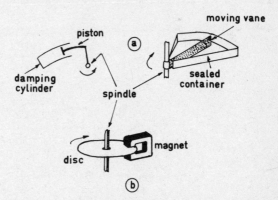

Figure 8.4

Part (a) of the figure shows two forms of *viscous* damping system in which a vane or piston attached to the moving part has to push a viscous medium, such as oil or air, out of the way when the pointer moves. This slows down the movement and design of the system and a suitable choice of medium is made to obtain optimum damping.

Electromagnetic damping is shown in fig. 8.4b. Here a disc attached to the moving spindle rotates through a magnetic field set up separately from the main instrument fields. Eddy currents are induced in the disc and produce their own small magnetic fields which react with the damping field so as to oppose motion. Movement still takes place but at a slower rate. The same principle of electromagnetic or eddy current damping is used in p.m.m.c. instruments but a different form is employed here. The moving coil is wound on a light metal former so that the eddy currents are induced in the coil former itself and not in a separate disc. The lateral twisting of a taut band introduces damping as well as a restoring force and there is usually no need for a further damping mechanism in this kind of instrument.

Use of the p.m.m.c. instrument in ammeters and voltmeters

For each basic p.m.m.c. movement (the name given to the magnet and coil assembly) there is a value of coil current to produce maximum deflection of the coil. This is called the full scale deflection or f.s.d. current, the corresponding coil p.d. being the f.s.d. voltage.

The basic movement may be incorporated into an instrument to measure current or voltage over any range of values, provided that the maximum current or voltage being measured does not allow a coil current or voltage greater than the f.s.d. value. To measure a current greater than the f.s.d. value, an additional parallel path must be provided to take the excess current so that a resistor called a *shunt* is connected across the movement coil. To measure a voltage in excess of the coil f.s.d. voltage an additional series resistor, called a *multiplier*, is used to drop the excess voltage (see fig. 8.5).

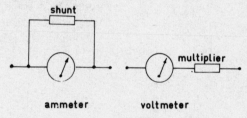

Figure 8.5

Example 8.1 A p.m.m.c. movement of resistance 50 Ω and f.s.d. current 10 mA is to be used as an ammeter reading current between zero and 1 A. What additional component is required for the movement?

The additional component is a resistor connected in parallel with the coil so that at full scale deflection it takes the *difference* between the coil current of 10 mA and the instrument (or meter) current of 1 A, i.e. 990 mA.

The f.s.d. p.d. across the coil = f.s.d. current × coil resistance

$$= 10 \times 10^{-3} \times 50 \text{ V}.$$

This p.d. is also across the shunt, since it is connected in parallel with the coil.

Shunt p.d. = 50×10^{-2} V; shunt current = 990×10^{-3} A

$$\text{Shunt resistance} = \frac{50 \times 10^{-2}}{990 \times 10^{-3}} = 5.05 \times 10^{-1} \, \Omega$$

The resistance of the shunt is very small and it must be capable of carrying 0.99 A at full scale deflection. In practice shunts are carefully and accurately wound of a conductor capable of carrying the required current without damage.

Example 8.2 A 60 Ω, 2 mA f.s.d. p.m.m.c. movement is to be used in a 0–100 V d.c. voltmeter. Calculate the value of any additional component and state how it must be connected.

At f.s.d. the p.d. across the movement is given by:

coil resistance × f.s.d. current = $2 \times 10^{-3} \times 60 = 120$ mV.

The f.s.d. instrument p.d. is 100 V, which means that 100 V − 20 mV must be dropped across the series multiplier resistance.

$$\text{Multiplier p.d.} = 100 \text{ V} - 20 \text{ mV} = 99.98 \text{ V}$$

Instrument current = 2 mA at f.s.d.

$$\text{Multiplier resistance} = \frac{99.98}{2 \times 10^{-3}} = 49.99 \text{ k}\Omega$$

The additional component is a resistance of 49 990 Ω connected in series with the p.m.m.c. movement.

Specific objectives

The expected learning outcome is that the student:
11.5 Describes with the aid of diagrams the principles of operation of an ohm-meter.
11.6 Uses an ohm-meter for the measurement of resistance.

The ohm-meter

Measurement of resistance may be made indirectly or directly. Indirect methods involve calculation (although this may be carried out automatically within the instrument in more sophisticated types), and measurements of voltage and current may first have to be taken. An instrument for direct reading of resistance, which displays the resistance value, is called an ohm-meter.

The simplest form of ohm-meter using a p.m.m.c. movement is shown in fig. 8.6. The circuit consists of a battery of terminal p.d. V volts applied to two resistors and the p.m.m.c. movement connected in series. One of the resistors, R, is of fixed value and is permanently connected within the instrument. The second resistor, shown as R_X, is the one of which the resistance is being measured and is connected across the ohm-meter terminals.

If the value of R_X is zero, i.e. the instrument terminals are short-circuited, the instrument current

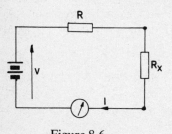

Figure 8.6

$$I = \frac{V}{R}$$

and R may be chosen so that this value of I is the f.s.d. value. This point on the scale, the f.s.d. point, is then marked 'zero ohms' showing that R_X is zero.

If the instrument terminals are left open-circuited the instrument current is zero and there is no deflection. This point on the scale is marked ∞ indicating that R_X is, for all intents and purposes, infinitely high. (It is in fact the resistance of the air between the terminals.)

$$\text{When } R_X \text{ is not zero, } I = \frac{V}{R + R_X}$$

and the current will be determined by the value of R_X (in addition to the values of R and V) so that the deflection will be somewhere between zero ($R_X = \infty$) and maximum ($R_X = 0$). Known values of R_X can be written on the scale when the instrument is calibrated so that in use, when R_X is not known, it may be read off directly.

Since for constant battery voltage $I \propto 1/(R + R_X)$

the scale is not linear (for this to be so I would have to be directly proportional to R_X alone), and tends to be cramped as the value of R_X increases.

'Set zero' controls may be incorporated into the instrument as shown in fig. 8.7.

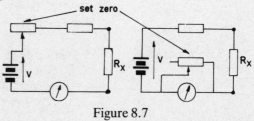

Figure 8.7

The 'set zero' control, which is adjusted prior to use with the instrument leads connected together, compensates for lead resistance, the effect of which is then removed, and for any variation in the battery e.m.f. between use.

Specific objectives

The expected learning outcome is that the student:
11.7 Explains the need for rectifier instruments.
11.8 States the frequency and waveform limitation inherent in moving-coil rectifier instruments.

Rectifier instruments

The p.m.m.c. movement responds usefully only to direct current. Alternating current in the coil would set up a coil magnetic field which would reverse its direction each time the coil current direction of flow were reversed and the deflecting force on the coil would act first in one direction and then the other. If the frequency of alternation is low the resultant deflection is an oscillation of the pointer or scale which can be observed. At common frequencies of 50 Hz and above the rate of change is too rapid for the deflection system to follow and the result is a vibration of the pointer or scale but no readable deflection.

To use a p.m.m.c. instrument to measure a.c. quantities the alternating coil current is first rectified as shown in fig. 8.8.

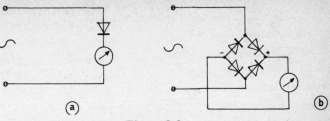

Figure 8.8

In part (a) of the figure a half-wave rectifier is used to give a pulse of current in one direction each full cycle. The full-wave rectifier shown in part (b) of the figure gives two pulses of current acting in the same direction during each full cycle of the circuit current.

In both cases the deflecting system cannot follow the coil current from zero to maximum to zero, etc. because of inertia but, since the deflection is now continually in the same direction, the final deflection is proportional to the *average* value of the current in the coil.

The ratio between the more commonly required r.m.s. value of an alternating current or voltage and its average value is called the form factor, which is constant provided that the waveform is regular and changing uniformly. For a pure sine wave, for example, the form factor is 1.11 and the scale of a rectifier p.m.m.c. instrument may be calibrated directly to read r.m.s. values by writing on the scale at any point the value obtained by multiplying the form factor by the average value of the quantity being measured which is causing the deflection at that point.

Once calibrated using a particular form factor the instrument is then, of course, only accurate, within the other limits of instrument accuracy, for that value of form factor. Measurement of alternating quantities having a waveform of different form factor will have an error due to the different form factor. Normally a rectifier instrument calibrated to read r.m.s. values is calibrated for sinusoidal waveforms. Any other kind of waveform not having a form factor of 1.11, including, of course, a sinusoidal waveform which has been distorted, either within the instrument or elsewhere, will not be measured within the same limits of accuracy as a pure sinusoid.

A second limitation of rectifier p.m.m.c. instruments is the *lower* frequency limit, below which the fluctuation of the pointer becomes visible and makes reading of the scale extremely difficult. The lower frequency figure is usually in the region of 20 Hz.

Specific objective

The expected learning outcome is that the student:
11.10 Uses a wattmeter.

The dynamometer wattmeter

A dynamometer instrument works on the same principle as the p.m.m.c. instrument. In this case, however, the permanent magnet is replaced by stationary coils and it is the reaction between the magnetic field set up by the fixed coils and the field set up by the moving coil which produces the deflection (see fig. 8.9).

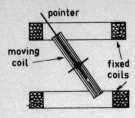

Figure 8.9

The instrument may be used as an ammeter or voltmeter by a suitable inter-connection of the coils but is more commonly used as a wattmeter to measure power in a load. In this case the load voltage and load current producing the power are fed to the instrument, the voltage to one coil (usually the moving coil) and the current to the others. The current coils may be connected in series or parallel depending upon the size of the load current. The voltage establishes a current in its coil, proportional to the voltage, and the reaction between the two fields, one produced by the load voltage, the other by the load current, causes a deflection proportional to the product of the two quantities.

Measurement of power using a separate voltmeter and ammeter is not straightforward since although these meters may measure r.m.s. values they take no account of power factor. The wattmeter does, since the deflection depends on the average value of the voltage–current product over a period of time. Since the deflection depends upon a product, the scale is a 'square law' scale, i.e. is non-linear.

Specific objectives

The expected learning outcome is that the student:
11.4 Calculates the power consumed in the instrument and the load.
11.6 Uses an ohm-meter for the measurement of resistance.
11.9 Uses multimeters correctly for the measurement of I and V in d.c. and a.c. circuits.
11.10 Uses a wattmeter.

Errors in measurement using electromechanical instruments

Any instrument used for measurement in an electrical circuit should not affect the circuit by its insertion. This is the ideal situation but, certainly with electromechanical instruments, it is not practically possible to attain. The main reason is that, unlike electronic instruments, this particular type draws current in order for it to function. Thus, a voltmeter will draw current and change current levels, an ammeter will have a p.d. across it and affect voltage levels and a wattmeter will absorb power itself and affect power levels if the instruments draw current from the circuit.

To reduce the error caused by the instrument a voltmeter should have as high a resistance as possible and require as small a current as possible for deflection. A characteristic commonly used to indicate the quality of a voltmeter in this respect is the 'ohms-per-volt' rating of the meter. This figure is obtained by dividing the meter resistance on a particular scale by the f.s.d. voltage on that scale. The higher the figure the better is the instrument.

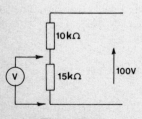

Figure 8.10

Example 8.3 A circuit consisting of a 15 kΩ resistor and a 10 kΩ resistor connected in series is connected across a 100 V d.c. supply (see fig. 8.10). Calculate the p.d. across the 15 kΩ resistor. A voltmeter is connected across the 15 kΩ resistor, the voltmeter range being 0–100 V. Calculate the p.d. across the 15 kΩ resistor with the voltmeter in circuit if its 'ohms-per-volt' rating is (a) 1 kΩ/V; (b) 22 kΩ/V.

The circuit total resistance is 15 kΩ + 10 kΩ, i.e. 25 kΩ, and the

$$\text{p.d. across the 15 k}\Omega\text{ resistor} = \frac{15 \times 10^3 \times 100}{25 \times 10^3} = 60 \text{ V}$$

(by potential division)

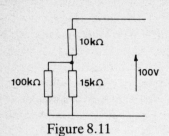

Figure 8.11

(a) If the voltmeter scale is 0–100 V and the 'ohms-per-volt' are 1 kΩ/V, the total voltmeter resistance on this scale is:

$$\text{f.s.d. voltage} \times \text{ohms/volt} = 100 \times 1 \text{ k}\Omega = 100 \text{ k}\Omega$$

This resistance is in parallel with the 15 kΩ resistor (fig. 8.11) and reduces the effective resistance to:

$$\frac{15 \times 100}{115} = 13.04 \text{ k}\Omega \text{ (product/sum)}$$

The new circuit resistance is therefore (10 + 13.04) kΩ, i.e. 23.04 kΩ

And the p.d. across the 13.04 kΩ resistance (by the potential division method) is:

$$\frac{13.04}{23.04} \times 100 \text{ i.e. } 56.6 \text{ V}$$

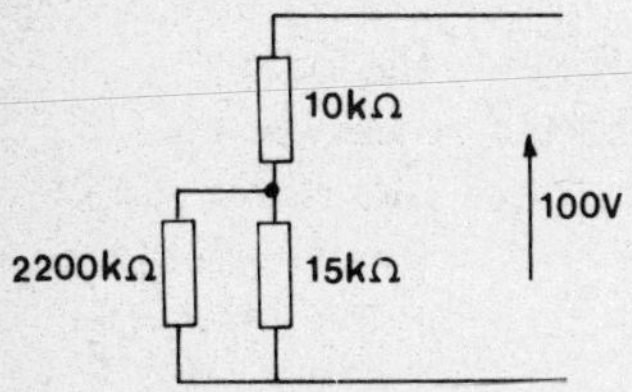

Figure 8.12

(b) When the voltmeter scale is 0–100 V and the 'ohms-per-volt' are 22 kΩ/V, the total voltmeter resistance in this scale is

$$100 \times 22 \text{ k}\Omega, \text{ i.e. } 2200 \text{ k}\Omega$$

This resistance is in parallel with the 15 kΩ resistor (fig. 8.12) and reduces the effective resistance to

$$\frac{2200 \times 15}{2215} \text{ k}\Omega, \text{ i.e. } 14.9 \text{ k}\Omega$$

The new circuit resistance is therefore 24.9 kΩ and the p.d. across the 14.9 kΩ resistance is given by

$$\frac{14.9}{24.9} \times 100 \text{ V, i.e. } 59.84 \text{ V,}$$

and we see that when no instrument is connected the p.d. is 60 V, when a 1 kΩ/V voltmeter is connected the p.d. is 56.6 V and when a 22 kΩ/V voltmeter is connected the p.d. is 59.84 V. The meter having the higher 'ohms/volt' rating causes the smaller error. Switching the meter to a higher range (0–200 V, 0–500 V etc.), if this is possible, increases the total meter resistance but reading the instrument becomes more difficult and a different kind of error is introduced.

An ammeter should have as low a resistance as possible so that the p.d. across it is as small as possible. A similar characteristic to the 'ohms/volt' figure for a voltmeter is not used for an ammeter but the meter with the lowest possible resistance should be chosen if a choice is available.

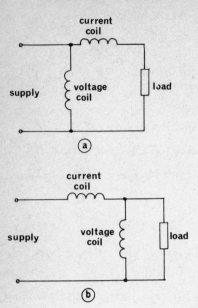

Figure 8.13

A wattmeter may be connected in circuit in one of two ways as shown in fig. 8.13. In part (a) of the figure, the voltage coil is reading the p.d. across the load *and* the current coil; in part (b) of the figure the current coil is reading the current in the load *and* the voltage coil.

In part (a) the current coil is reading the correct load current but not in part (b); in part (b) the voltage coil is reading the correct load voltage but not in part (a). There is not a method of connection which allows both coils to read correctly. Which method of connection is used depends largely on the relative error introduced by each coil for a particular application. Error correction is made either by calculation or in some instruments by the use of internal compensating coils to reduce the effect of the additional voltage or additional current caused by the connection method.

When using a wattmeter, the maker's instructions should be read and followed carefully; guidance on the error introduced by connection is usually given with each meter. The following examples should be studied carefully.

Example 8.4 A wattmeter having a voltage-coil resistance of 10 kΩ and total current-coil resistance of 5 Ω is connected to measure the power in a 100 Ω resistive load. The supply p.d. is 250 V. Calculate the wattmeter reading and the true power in the load:

(a) when the voltage coil is connected across the load in series with the current coil, i.e. the voltage coil is across the supply.

(b) when the current coil is connected in series with the parallel combination of the load and the voltage coil.

What is the power in the load when the wattmeter is not connected?

(a) The voltage-coil p.d. is 250 V. In the circuit branch containing the load, the total resistance across the supply is the sum of the load resistance and the current-coil resistance, i.e. 105 Ω, so

$$\text{the load current} = \frac{250}{105} = 2.475 \text{ A}$$

$$\text{Wattmeter reading} = 250 \times 2.38 = 595.24 \text{ W}$$

$$\text{True power in load} = \text{current}^2 \times \text{resistance}$$
$$= 2.38^2 \times 100 = 566.89 \text{ W}$$

(b) The resistance of the parallel combination of the 10 kΩ voltage coil and the 100 Ω load is

$$\frac{10\,000 \times 100}{10\,100}, \text{ i.e. } 99 \text{ }\Omega$$

Total circuit resistance = 99 + current coil resistance
= 104 Ω

Current drawn from supply through the current coil

$$= \frac{250}{104} = 2.4 \text{ A}$$

P.D. across the current coil = 2.4 × 5 = 12 V

Therefore, p.d. across voltage coil = 250 − 12 = 238

Wattmeter reading = 238 × 2.4 = 571.2 W

True power in load = (load voltage)^2/resistance

$$= \frac{238}{100} = 566.44 \text{ W}$$

When no meter is connected:

Power in load = (supply voltage)^2/resistance = $\frac{250^2}{100}$ = 625 W

And we see that the true power with no meter connected, 625 W, may be reduced to 566.89 W read as 595.24 W, or 566.44 W read as 571.2 W, depending on the connection mode. Mode (a) is clearly the better here since the wattmeter reading corresponds more closely to the actual power when no meter is connected.

As a guide to which is the better connection for a particular application, if the voltage-coil resistance is much larger than the load resistance, method (b) should be used (current coil taking both voltage-coil and load current) and, if the current-coil resistance is much smaller than the load resistance, method (a) should be used (voltage-coil p.d. is the sum of the current-coil p.d. and the load p.d.)

The idea in general is to cause least disturbance to the circuit.

Example 8.5 A wattmeter having a voltage-coil resistance of 5 kΩ and a current-coil resistance of 0.8 Ω is connected to measure the power in a resistive load connected to a 200 V d.c. supply. When the wattmeter is connected so that the voltage coil is in parallel with the series combination of the current coil and the load, the reading is 500 W. Calculate the true power in the load:
(a) when the wattmeter is connected; (b) when the wattmeter is not connected.

(a) Voltage-coil p.d. = 200 V; wattmeter reading = 500 W

Current in the current coil (and load) = $\frac{500}{200}$ = 2.5 A

Current-coil p.d. = 2.5 × 0.8 = 2 V

Load p.d. = 200−2 = 198 V

Thus, true power = 198 × 2.5 = 495 W

Alternatively, and more simply, power loss in current coil

$$= 2.5^2 \times 0.8 = 5 \text{ W}$$

and true power = wattmeter reading minus power loss

$$= 500−5 = 495 \text{ W}$$

(b) From part (a), load p.d. = 198 V; load current = 2.5 A

Load resistance = $\frac{198}{2.5}$ = 79.2 Ω

and true power when the wattmeter is not connected

$$= \text{(supply p.d.)}^2/\text{load resistance} = \frac{200^2}{79.2} = 505 \text{ W}.$$

Example 8.6 The power in a load connected to a 250 V d.c. supply is measured by a wattmeter connected so that the voltage coil is in parallel with the load. The wattmeter voltage-coil resistance is 10 kΩ and its current-coil resistance is 1 Ω. The wattmeter reading is 800 W. Calculate the true power in the load if an ammeter of negligible resistance connected in series with the current coil reads 3.25 A.

Wattmeter current = 3.25 A; wattmeter reading = 800 W

$$\text{Thus, voltage-coil p.d.} = \frac{800}{3.25} = 246.15 \text{ V}$$

$$\text{Voltage-coil current} = \frac{246.15}{10\,000} = 0.0246 \text{ A}$$

$$\text{Load current} = 3.25 - 0.0246 = 3.2254 \text{ A}$$

$$\text{True power} = \text{load p.d.} \times \text{load current}$$
$$= 246.15 \times 3.2254 = 793.9 \text{ W}$$

Alternatively, and more simply:

$$\text{Voltage-coil p.d.} = 246.15$$

$$\text{Power loss in voltage coil} = \frac{246.15^2}{10\,000} = 6.1 \text{ W}$$

True power = wattmeter reading minus power loss in voltage coil
$$= 800 - 6.1 = 793.9 \text{ W}$$

In general we note that:
(1) When the voltage coil is across the supply (i.e. across the series combination of current coil and load),

True power = wattmeter reading minus power loss in current coil.

(2) When the voltage coil is across the load (i.e. the current coil carries the load current and the voltage-coil current).

True power = wattmeter reading minus power loss in voltage coil.

Errors of instruments due to causes other than connection in circuit depend on the instrument type. P.M.M.C. instruments suffer from weakening of the magnetic field as the instrument ages but other errors are relatively small. Errors in rectifier-type instruments were considered earlier. Moving-iron meters are subject to errors caused by hysteresis of the magnetic circuit and, since they usually work at higher power levels, may be affected by increased temperature in use. Dynamometer wattmeters suffer from an error due to the inductive reactance of the coils which may affect the effective power factor when the instrument is connected. The reading is usually multiplied by a 'correction factor' given by the manufacturer.

Care should be taken when reading instruments employing a pointer and scale. If the meter is read at an angle, i.e. the line of sight is not perpendicular to the scale, errors of *parallax* occur. To reduce this, most instruments have a mirror mounted behind the pointer and to read the meter correctly the pointer and its image are 'lined up' with the eye before reading the scale (see fig. 8.14).

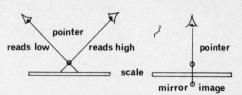

Figure 8.14

When using multipurpose meters, the user should ensure:

(1) The instrument is in the correct *mode*, i.e. voltmeter, ammeter, etc. Connecting a multimeter set to read current in parallel with a supply will almost certainly damage the meter.

(2) The instrument is switched to the *highest* range when starting measurement. If the deflection is too small the range may be progressively reduced to obtain a suitable deflection.

(3) The instrument is *never* left in the ohm-meter mode. Apart from the risk of touching leads running down the internal battery, the next user may connect the meter to a circuit without checking the mode. The ohm-meter mode applies a p.d. to the circuit and non-checking may result in damage to the circuit or to the meter or to both.

Specific objective

The expected learning outcome is that the student:
11.11 *Uses d.m.m., c.r.o. and electronic voltmeters in direct and indirect measurement of a.c. and d.c. voltages and the c.r.o. to measure period and frequency.*

The cathode ray oscilloscope

The cathode ray oscilloscope (c.r.o.) is one of the most versatile of electronic instruments since it is capable not only of measurement of a quantity but also of display of a graph of the quantity plotted against time. More sophisticated instruments can capture extremely fast changes and store the information concerning the rate of change and values reached of the changing quantity. A typical modern c.r.o. is shown in the photograph.

The heart of the c.r.o. is the cathode ray tube (c.r.t.), an electrostatic form of which is shown diagrammatically in fig. 8.15. The principle of the c.r.t. is that moving electrons strike a specially coated fluorescent screen, which glows at the point where the electrons strike.

The tube consists basically of an *electron gun* assembly, a *focussing* system and a *deflection* system linked together and sealed in an evacuated glass tube, the front of which, the screen, provides the

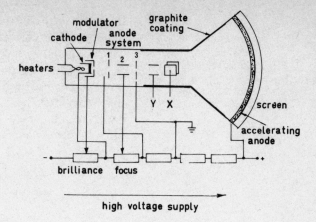

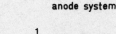

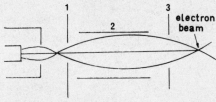

Figure 8.15

means of *display* of the waveforms of the quantity being examined.

The electron gun contains a heater filament, a cathode, which is coated with specially emissive material and when heated emits electrons, and a modulator, which is disc-shaped in cross-section, the disc containing a hole through which the electrons are drawn when the c.r.t. is being used for display. The modulator is held at a negative potential with respect to the cathode and controls the number of electrons passing through per second and eventually reaching the screen. The brightness of the glow at the screen depends upon this number so that the tube brightness control is connected to the modulator.

The electrons are attracted from the gun by a system of positive anodes, some of which are used to focus the electron flow into a beam so that a fine point of light appears at the screen. The anode system consists of two or more discs and cylinders which are held at suitable potentials to affect the electron flow, as shown in the figure.

Once formed into a beam the electron flow may be deflected vertically or horizontally so that the light spot moves up and down or from side to side on the screen. Using both vertical and horizontal deflection simultaneously the spot may be placed anywhere on the screen as desired.

In the electrostatic c.r.t., which is the type most commonly used in the c.r.o., deflection is achieved using electric fields set up by voltages applied to two sets of plates, one set, called the Y-plates, producing a vertical field and the other set, called the X-plates, producing a horizontal field. Deflection may also be produced using electromagnetic means, magnetic fields produced by coils, and this method is used in cathode ray tubes employed for large displays and in television and radar equipment.

To obtain maximum brightness at the screen the electrons must be moving at as high a speed as possible immediately before striking the screen. To obtain maximum deflection per volt of the deflection p.d., however, the electrons should be moving as slowly as possible through the deflection system. To solve this problem the initial speed of the beam, caused by the attraction of the (positive) anode focussing system is kept low and the electrons are further accelerated *after* passing through the deflection system by a final post-deflection anode situated close to the screen. This usually consists of a ring of conductive material placed on the inside of the tube.

C.R.T. voltages are relatively high, of the order of kilovolts, and great care must be taken when maintenance or other work is carried out on a live tube.

The main controls of the c.r.o., then, include a brightness or 'intensity' control, which alters the modulator potential with respect to the cathode, a focus control, which affects the relative anode potentials in the focusing system and thus shapes the beam, and a deflection control which controls the position of the light spot at the screen.

It is the use of the deflection control which enables the c.r.o. to display and measure waveforms of quantities. Any quantity can be measured and displayed provided that a signal voltage derived from, and proportional to, the quantity is available.

Waveform display

In the absence of a signal to either X- or Y-plates, the c.r.o. displays a spot of light at the screen, the position of which is determined by the direct voltages applied to the plates. The electron beam tends to be attracted to the more positive of the two plates in each of the sets as it passes through and it is this attraction which deflects the beam. Altering the relative potential of one plate with respect to the other changes the amount of deflection and thus determines the final position of the spot.

Control of the static or 'no signal' position of the spot is achieved by controls which may be marked 'vertical shift', 'Y-shift', 'horizontal shift', 'X-shift' or 'position' with an appropriate arrow indicating vertical or horizontal direction, the actual label being determined by the manufacturer. These controls alter the direct voltage applied to the X- and Y-plates.

If now an alternating voltage (in addition to the static direct voltage) is applied to the Y-plates only, the spot will move up and down the screen in a vertical line as one Y-plate becomes progressively more or less positive with respect to the other. The coating on most c.r.o. screens glows for a little while after the electrons strike it, the characteristic being called screen 'persistence', and because of this we do in fact see a continuous vertical line when an alternating voltage is applied to the Y-plates only. Similarly if an alternating voltage is applied to the X-plates only we would see a continuous horizontal line.

To display a voltage waveform, the signal to be displayed is applied to the Y-plates and at the same time the spot is deflected horizontally from left to right by a suitable voltage applied to the X-plates. As the spot moves to the right the voltage on the Y-plates is changing according to the instantaneous value of the signal so that at any particular time the position of the spot is determined vertically by the signal and horizontally by the time-varying voltage applied to the X-plates (see fig. 8.16).

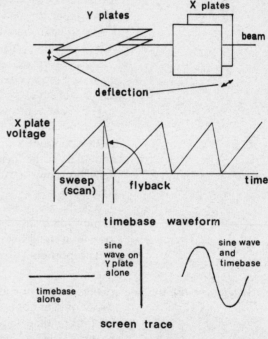

Figure 8.16

The time-varying voltage applied to the X-plates is called the *timebase* or *sweep* voltage. As shown in the figure the voltage rises linearly from zero to a maximum value and is then suddenly reduced to zero, when the process is started again. This variation of the X-plate voltage causes the spot to move horizontally to the right-hand side of the screen as the voltage rises and then return quickly to the left-hand side as the voltage is rapidly returned to zero. The wave is called a *saw-tooth* wave from its shape. The quick return period is called the *flyback* and usually during the flyback period, which is very small, a negative pulse of voltage is applied to the tube modulator to cut off the electron beam so that the flyback trace does not appear at the screen. The slower period of the waveform in which the signal is traced is called the *scan period*.

By adjusting the period of the timebase voltage the number of cycles of the signal waveform shown on the screen can be altered from a fraction of one cycle to as many as are required. If too many are shown, of course, the screen shows a rectangle with individual cycles being indistinguishable from each other.

Example 8.7 Calculate the number of cycles of a 2 kHz waveform shown on the screen of a c.r.o. when the timebase scan period is adjusted to 10 ms. What would the period need to be to show one cycle?

The periodic time of a regular alternating waveform is equal to the reciprocal of the frequency so that for a 2 kHz waveform,

$$\text{Periodic time} = \frac{1}{2 \times 10^3} \text{ s} = 0.5 \text{ ms}$$

In a scan period of length 10 ms therefore the number of cycles shown

$$\text{is equal to } \frac{10 \times 10^{-3}}{0.5 \times 10^{-3}}, \text{ i.e. } 20$$

Twenty cycles of the 2 kHz signal waveform would be shown during a scan period of 10 ms. To show one cycle of signal, the scan period must equal the periodic time of the signal, i.e. 0.5 ms.

Example 8.8 The timebase voltage of a c.r.o. is adjusted so that the scan period is 100 ms. The number of cycles of a 50 Hz waveform shown on the screen is:

   A. 2; B. 0.5; C. 0.2; D. 5.

A. To obtain this answer the scan period (time) appears to have been divided by signal frequency (cycles). In the first place both quantities should have the same units to give a pure number as the answer and secondly, the principle employed in getting this answer is quite wrong and gives a meaningless figure.

B. To obtain this answer the signal frequency has been divided by scan period. Comments as under 'A' apply again.

C. Here there is not an error from a unit point of view, since it appears that the periodic time of the signal (1/50 s or 20 ms) has been divided by the scan period (100 ms). This does however give the reciprocal of the correct answer.

D. The number of cycles of signal waveform shown is given by

$$\frac{\text{scan period}}{\text{period of one cycle of signal}}$$

and in this case equals 100 ms divided by 20 ms. The answer of 5 cycles is therefore correct.

To obtain a stationary trace on the screen it is necessary for the scan period to start at the same relative point on the signal waveform each time. If the scan period were to start later, for example, i.e. further to the right of the signal waveform, each time, the trace appears to move from right to left. If it starts earlier the trace movement is from left to right.

To ensure this does not happen the timebase voltage is *synchronised* with the signal using a pulse derived from the signal to trigger the circuit generating the timebase waveform. The c.r.o.

normally therefore has additional controls for this marked SYNC or TRIGGER.

Sometimes a signal consists of a combined a.c. and d.c. voltage and the trace then consists of an alternating waveform having its vertical position on the screen determined by the d.c. voltage of the signal (as well as the c.r.o. vertical shift controls). If this is not desired the d.c. level may be removed by inserting a capacitor in the input line to the Y-plates. This is done within the instrument and is controlled by a switch marked AC/DC on the outside of the c.r.o.

The vertical size of the trace depends upon the magnitude of the signal. In order that this might be measured by examination of the trace, the Y-plate controls are calibrated in terms of signal voltage per unit height of trace. The control usually consists of a switch showing volts/cm, the figure being different for each switch position e.g. 0.5 V/cm, 1 V/cm, 5 V/cm, and so on.

To obtain the value of the signal voltage at any point in the waveform, the height of the trace above a centre zero line at the point is multiplied by the calibration figure shown at the Y-plate control. For example if the Y-plate control setting is 0.2 V/cm and the height is 2 cm the signal voltage is 2×0.2, i.e. 0.4 V.

The c.r.o. can therefore be used to measure voltage and display waveform. To assist in measuring the vertical (or horizontal) length a grid is inserted at the front of the screen. This grid is called the *graticule*.

By calibration of the X-controls in terms of time so that the scan period is shown at each setting, the frequency of a signal can be determined. If the controls are altered so that an exact number of cycles are shown, the calculation is, of course, made much easier.

For example, if five complete cycles are shown in a scan period of 1 ms the periodic time of the signal (the time per cycle) is 1/5 ms, i.e. 0.2 ms

$$\text{and the frequency is } \frac{1}{0.2 \times 10^{-3}}, \text{ i.e. 5 kHz}$$

The c.r.o. may also be used in more sophisticated methods of frequency measurement.

Current may be measured and waveforms displayed by the use of a standard fixed resistor of known value. If the current under examination is passed through the resistor, the p.d. developed across it can be displayed and measured and hence the value of the current can be determined using Ohm's Law. The p.d. waveform is, of course, identical to the current waveform.

The photograph is of a fairly typical laboratory instrument. This c.r.o. is capable of showing two traces at the same time and has two Y-input channels (CH1 and CH2). The inputs are on the left-hand side of the instrument. To the left of the screen are situated the Y-plate voltage sensitivity controls for each channel, the centre knobs controlling the vertical shift (the no-signal vertical position of the spot). The available volts per division (meaning one division of the graticule) are clearly seen to range from 5 mV to 20 V.

The timebase control is on the right-hand side, the centre knob being a fine control to adjust the trace horizontally. Here we have time per division ranging from 0.2 μs to 0.2 s. The horizontal shift control is situated just below the timebase current.

Focus, intensity and power controls are shown underneath the screen. This particular c.r.o. has an additional feature which is extremely useful, the 'beamfinder'. When setting up a c.r.o. for use the trace is sometimes found to be off the screen due to the settings left by the last user. The 'beamfinder' button brings the trace to the screen centre and by noting the direction from which it came the X- and Y-shift controls can be adjusted accordingly.

Other controls include the channel AC/DC controls on the left and TRIGGER controls on the right.

One final point concerning the use of the c.r.o. is that the intensity control should not be turned up too high and that a bright stationary trace should not be left on the screen for long periods; otherwise 'screen burn' may occur which permanently marks the screen.

Specific objectives

The expected learning outcome is that the student:
11.12 Describes the principle of a null method of measurement.
11.13 Describes with the aid of a diagram the principle of the Wheatstone bridge.

Null and bridge methods of measurement

Other than using instruments and meters and the c.r.o., measurement of electrical quantities may also be carried out, usually to a high degree of accuracy, by null and bridge methods.

The commonest null method employs the simple d.c. potentiometer and a variety of more sophisticated methods are based on the same basic principle. The principle, simply, is that the voltage being measured is set against a known voltage and the known voltage adjusted until a zero reading, called a *null*, is obtained on an extremely sensitive centre-zero ammeter, called a *galvanometer*. When the null is obtained the value of the unknown voltage may be obtained by reference to the known voltage.

The d.c. potentiometer

Fig. 8.17 shows the basic layout and circuit of a simple d.c. potentiometer. It consists of a slide-wire and a moving contact, the slide-wire being mounted against a rule so that the position of the sliding contact relative to one end of the slide-wire may be read off easily.

A voltage source, E_1, and variable resistor R, are connected in series with the slide wire, a separate voltage source, E_2, in series with a galvanometer, G, being connected between one end of the slide-wire and the sliding contact.

In the galvanometer circuit there are two voltages: one is E_2, the other is the voltage across the slide-wire and it is the *difference* between these voltages which determines the direction of current flow in the galvanometer. The slide-wire voltage may be varied by moving the sliding contact so that it may be smaller than E_2, larger than E_2 or equal to E_2. If it is smaller than E_2 then current will flow

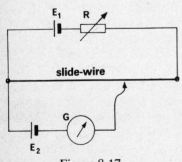
Figure 8.17

clockwise in the circuit; if it is larger than E_2, current flow is anticlockwise; if it is equal no current flows and the *null* point is reached.

To use the potentiometer it must be possible to calculate the slide-wire voltage at the null point. To do this the potentiometer is calibrated using a standard cell. A standard cell is one which is guaranteed to provide a specific value of e.m.f. within very close limits of accuracy. One commonly used is the Weston cadmium cell having an e.m.f. of 1.0186 V.

The standard cell is connected in circuit in the E_2 position and the sliding contact adjusted until a null point is reached. At this point the length of the slide-wire is read off the rule which is mounted beneath the wire. The wire itself is uniform throughout its length (having the same cross-sectional area and resistive properties) so that the voltage drop per unit length for this setting of E_1 and the variable resistor can be calculated.

If the standard cell e.m.f. is denoted by E_s (volts) and the length of slide-wire (centimetres) by ℓ_1, then the voltage drop per unit length is

$$\frac{E_s}{\ell_1} \text{ volts/cm}$$

The standard cell is now removed and *with E_1 and the setting of R unaltered*, the unknown voltage is connected in the E_2 position. The sliding contact is moved to the new null point and the length of slide-wire read off. At the new null point the slide-wire voltage is equal to the unknown voltage.

If the new slide-wire length is denoted by ℓ_2 (centimetres),

Slide-wire voltage = length of slide-wire × volts/cm = $\ell_2 \times \dfrac{E_s}{\ell_1}$

and, denoting the unknown voltage by E_X,

Slide-wire voltage = $E_X = \dfrac{\ell_2}{\ell_1} E_s$

This is the basic potentiometer equation.

It should be noted that in using the potentiometer, the value of E_1 and the setting of the variable resistor are chosen to give a reasonable length of slide-wire at the first null (with the standard cell in circuit) and *must not then be altered* when the unknown voltage is connected. If it is changed the volts/cm figure is changed.

The simple potentiometer can be made effectively direct-reading by arranging for the length of slide-wire at the first null to have the same significant figures as the standard cell e.m.f. For example, if the standard cell e.m.f. is 1.0186 V, the length may be arranged to be 101.86 cm, i.e. a numerical value 100 times the numerical value of the standard e.m.f. The second length is then read off and divided by 100 to give the numerical value of the unknown e.m.f.

The volts/cm in this case would be $\dfrac{1.0186}{101.86}$, i.e. 0.01

and this is the figure ($\frac{1}{100}$) which is multiplied by the second length to give the unknown voltage.

To calibrate in this manner, the standard cell is connected and the sliding contact set at the desired length (101.86 cm in the example given). The variable resistor R is then adjusted to give a null. At this point the standard cell e.m.f. is equal to the voltage across the set length and the required volts/cm figure is set. The setting of R and the value of E_1 are then left unaltered and E_X is connected in circuit.

A distinct advantage of the simple potentiometer method is that at the null point no current is drawn from the unknown voltage source and the value obtained is the true e.m.f. not the terminal p.d. Voltmeters which draw current cannot measure the open-circuit e.m.f.

As with the c.r.o., provided that a voltage can be obtained from the quantity, any quantity can be measured using the potentiometer. Current may be measured by measuring the p.d. across a standard resistor when the current is flowing in the resistor. A derivation of the potentiometer method used for measuring resistance, called a *bridge*, is considered next.

The Wheatstone bridge

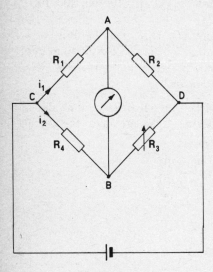

Figure 8.18

Fig. 8.18 shows a basic Wheatstone bridge circuit for the measurement of resistance. The circuit consists of four resistors, one of which is unknown, one of which is variable, the other two being fixed, a voltage source and a galvanometer. Current drawn from the supply flows via R_1 and R_2 and via R_3 and R_4 and also in the galvanometer when the bridge is not at a null point.

The magnitude and direction of the galvanometer current depends upon the magnitude of the potential at A with respect to B and this in turn depends upon the values of the resistors and the current flowing in them. If point A is at a higher potential than B, current will flow from A to B; if point A is at a lower potential than B, current will flow from B to A. When the potential difference between A and B is zero no current flows in the galvanometer, a null point is reached and the bridge is said to be *balanced*.

Balance is controlled by varying the value of resistor R_3 in the circuit shown and variation over its range will give firstly a galvanometer deflection in one direction, no deflection, then a deflection in the opposite direction.

At balance, if i_1 represents the current in R_1 and R_2, and i_2 represents the current flowing in R_3 and R_4 since there is no p.d. between A and B, these points must be at the same potential with respect to any other point so that:

$$\text{p.d. between A and C} = \text{p.d. between B and C}$$

$$\text{and } i_1 R_1 = i_2 R_4$$

also, p.d. between A and D = p.d. between B and D

$$\text{and } i_1 R_2 = i_2 R_3$$

Dividing the first equation by the other:

$$\frac{R_1}{R_2} = \frac{R_4}{R_3}, \text{ which is the balance equation of the bridge.}$$

This may be rearranged to give $R_4 = R_3 \times \frac{R_1}{R_2}$.

and if R_4 is the unknown resistor, this equation can be used to give the value of the unknown resistance in terms of the known resistances R_3, R_2 and R_1.

The known resistors R_1, R_2 are called the *ratio arms* of the bridge and an appropriate value of the ratio R_1/R_2 may be chosen to make the bridge effectively or actually direct reading.

If $\frac{R_1}{R_2} = 1$, for example, $R_4 = R_3$ at balance and the value of R_4 can be read off directly.

By choosing an appropriate ratio, the value of R_4 may be measured very accurately indeed. Suppose R_4 is relatively small at, say, 1.021 Ω, by arranging the ratio R_1/R_2 to be 1/1000 the bridge would balance at $R_3 = 1021$ Ω, a value quite easily obtainable using a standard decade resistance box (a variable resistor with switch resistors in sets, each set a decade, e.g. 0–10, 10–100, 100–1000, etc.).

If R_4 is relatively large then the ratio R_1/R_2 may be unity or at much higher values as appropriate. For example, if R_4 is 10.67 MΩ and R_1/R_2 is chosen as 1000/1, the bridge would balance at $R_3 = 10\,670$ Ω, again a value easily obtainable.

Example 8.9 The circuit of fig. 8.18 is used to measure the value of an unknown resistor in the R_4 position. Calculate R_4 for the following values of the other resistors:

(a) $R_3 = 4714$ Ω, $R_2 = 1000$ Ω, $R_1 = 10$ Ω.
(b) $R_3 = 101$ Ω, $R_2 = 100$ Ω, $R_1 = 10$ Ω.
(c) $R_3 = 9875$ Ω, $R_2 = 100$ Ω, $R_1 = 100$ Ω

(a) Ratio $\frac{R_1}{R_2} = \frac{10}{1000}$, i.e. $\frac{1}{100}$

$$R_4 = \frac{R_3}{100} = 47.14 \text{ Ω}$$

(b) Ratio $\frac{R_1}{R_2} = \frac{10}{100} = \frac{1}{10}$

$$R_4 = 10.1 \text{ Ω}$$

(c) Ratio $\frac{R_1}{R_2} = \frac{100}{100} = 1$

$$R_4 = R_3 = 9875 \text{ Ω}$$

Summary

There are two main kinds of electrical instrument, electromechanical and electronic. The first uses electromagnetic measurement and mechanical display; the second uses electronic measurement and display. There are also hybrid instruments.

The essential features of an electromechanical instrument are deflection, control and damping. The quantity being measured causes a deflection of the indicating part and a control force to counteract the deflecting force is set up. When the two are equal the indicator stops moving. The degree of damping controls how quickly the instrument settles down into the steady state.

Deflection may be caused by magnetic attraction or repulsion of iron pieces as in the moving-iron instrument or by the forces set up between a magnetic field and current-carrying conductor as in the permanent-magnet moving-coil (p.m.m.c.) or d'Arsonval instrument. The basic instrument movement gives full scale deflection (f.s.d.) for a certain voltage and current. To increase the instrument range for use as a voltmeter a series resistance called a multiplier is added; to increase the range for use as an ammeter a parallel resistance called a shunt is added.

P.M.M.C. instruments are used in ammeters, voltmeters, ohm-meters and in a wide variety of instruments measuring other quantities. They require direct current for operation and if used in a.c. circuits the current must first be rectified. The scale deflection in a p.m.m.c. instrument is linear with equally spaced divisions on the scale. The accuracy of a p.m.m.c. instrument used in a.c. circuits is affected by the purity of the waveform of the quantity being measured.

The main use of moving-iron instruments is the measurement of current and voltage especially in a.c. circuits. The scale is non-linear or 'square-law', being cramped at one end.

The dynamometer instrument uses two sets of coils, one of which is able to move and is used as the indicator. It may be used as an ammeter or voltmeter but is most commonly used as a wattmeter. When used as a wattmeter one set of fixed coils is used for current, the moving coil being used for voltage, the combination producing a deflection proportional to power.

There are two methods of wattmeter connection, both of which are subject to reading error due to power loss in the coils. When the voltage coil is connected across the supply, the current-coil power should be deducted from the wattmeter reading to give the true power. When the voltage coil is connected across the load the true power is the wattmeter reading less the voltage-coil power.

When using instruments the following points should be noted:
(1) Voltmeters are connected in parallel with the load; a good-quality voltmeter has a very high resistance compared with the load, a measure of this being the 'ohms-per-volt' rating of the meter. Total meter resistance is ohms-per-volt × f.s.d. voltage for the range to which the meter is switched.
(2) Ammeters should be connected in series with the load and should have as low a resistance as possible.

(3) Ohm-meters should never be connected to live circuits and should be zeroed before use to compensate for lead resistance. Multimeters should not be left in the ohm-meter mode.
(4) When using a multimeter it should be in the correct mode and measurement is commenced on the highest possible range.

The cathode ray oscilloscope (c.r.o.) uses the movement of a light spot, caused by electrons impinging on a fluorescent screen, to both indicate and measure. The electron beam may be moved sideways (the X-direction) or vertically (the Y-direction) by applying voltages to sets of plates, called X- and Y-plates respectively. By using a timebase voltage on the X-plates the c.r.o. will trace a waveform of a voltage applied to the Y-plates. The c.r.o. should never be left showing a very bright stationary trace on the screen as this may cause damage.

Null methods of measurement include the d.c. potentiometer, in which an unknown e.m.f. is compared with a known e.m.f., and the Wheatstone bridge, in which adjustment of a variable resistor causes the p.d. across opposite arms of the bridge to become equal. A null is reached in both methods when a special sensitive centre-zero ammeter, called a galvanometer, takes zero current.

EXERCISE 8

1. A 10 mA f.s.d., 25 Ω p.m.m.c. movement is to be used as a 0–100 V d.c. voltmeter. Calculate the value of the multiplier resistance.

2. Calculate the value of shunt resistance required for a 0–1 A ammeter using a 10 mA f.s.d., 50 Ω p.m.m.c. movement.

3. A 0–100 V d.c. p.m.m.c. voltmeter is connected across a 1 kΩ resistor which forms part of a separate circuit. The current flowing into the resistor/voltmeter combination is 99 mA, the p.d. across it being 98.5 V. Calculate the ohms/volt rating of the voltmeter.

4. A 1 mA f.s.d. moving-coil instrument of coil resistance 40 Ω is used as a 0–250 V d.c. voltmeter. Calculate the ohms/volt rating of the voltmeter.

5. A 0–5 A d.c. ammeter using a p.m.m.c. movement of coil resistance 2Ω and f.s.d. current 10 mA is connected in series with a 10 Ω resistor across a 50 V d.c. supply. Determine the change in circuit current caused by the inclusion of the ammeter.

6. A circuit consisting of a 20 kΩ resistor connected in series with a 30 kΩ resistor is connected across a 50 V d.c. supply. A 0–50 V d.c. voltmeter having an ohms/volt rating of 10 kΩ/V is connected across the 20 kΩ resistor. Calculate the p.d. across the resistor before and after connection of the voltmeter.

7. A 1 mA, 25 Ω p.m.m.c. movement is to be used in a multimeter to read the following ranges of d.c. current and voltage: 0–1 V; 0–10 V; 0–100 V; 0–1 A; 0–5 A; 0–20 A. Calculate the value of the shunt or multiplier resistance for each range.

8. A dynamometer wattmeter of voltage-coil resistance 10 kΩ and current-coil resistance 0.8 Ω is connected to read the power in a 150 Ω resistor connected across a 250 V d.c. supply. Calculate the wattmeter reading and the true power in the load for both methods of wattmeter connection.

9. The power in a resistive load connected to a 150 V d.c. supply is read as 200 W by a wattmeter connected with its voltage coil in parallel with the supply. The resistance of the wattmeter voltage and current coils is 10 kΩ and 1 Ω respectively. Calculate the true power in the load:
   (a) when the wattmeter is connected; (b) when the wattmeter is not connected.

10. The voltage coil of a wattmeter has a resistance of 20 kΩ and is connected across a resistive load connected to a 250 V d.c. supply. The current-coil resistance is 0.8 Ω. The wattmeter reading is 900 W; calculate the true power in the load.

11. The timebase scan period of a c.r.o. is 20 ms. Calculate the number of cycles shown of a waveform of frequency:
   (a) 1 kHz; (b) 10 kHz; (c) 1 MHz.

12. A c.r.o. shows five complete cycles of a waveform when its scan period is 10 ms. The frequency of the waveform is:
   A. 500 Hz; B. 200 Hz; C. 2 Hz; D. 0.5 Hz.

13. A rectifier instrument connected to a sinusoidal a.c. supply reads 45 V. Calculate the length of the trace between peaks when the supply is connected to the Y-plates of a c.r.o. set to a deflection sensitivity of 10 V/cm.

14. A d.c. potentiometer is calibrated using a standard cell of e.m.f. 1.0186 V, the slide-wire length being 80 cm. When a cell of unknown e.m.f. is connected to the circuit the null point is obtained at a slide-wire length of 100 cm. Calculate the unknown e.m.f.

15. A d.c. potentiometer is calibrated using a standard cell of e.m.f. 1.0186 V so that the slide-wire length is 101.86 cm. What would be the slide-wire length if a cell of e.m.f. 1.95 V were connected to the potentiometer?

16. A Wheatstone bridge has the following values of resistance at balance: $R_1 = 100$ Ω; $R_2 = 1000$ Ω; $R_3 = 572$ Ω. Determine the value of R_4.

17. In the circuit of fig. 8.18 the resistance of R_4 is known to be 871 Ω. Balance is obtained when R_3 has the value of 8710 Ω. What is the ratio R_1/R_2?

18. Suggest suitable values of the ratio R_1/R_2 in the circuit of fig. 8.18 if R_3 is a decade box with four decade settings 0–10 Ω, 10–100 Ω, 100–1000 Ω and 1000–10 000 Ω, and R_4 has a value which:
   (a) lies between 1 Ω and 2 Ω; (b) lies between 1000 Ω and 2000 Ω; (c) lies between 1 MΩ and 2 MΩ.
The setting must be such that the resistance of R_4 is obtained to four significant figures.

Possible marks

SELF-ASSESSMENT EXERCISE 8

1. Calculate the value of multiplier required when a 1 mA f.s.d., 15 Ω p.m.m.c. movement is used in a 0–100 V voltmeter. (3)

2. A 10 mA, 25 Ω p.m.m.c. movement is used in a 0–1 A ammeter. Calculate the shunt resistance. (3)

3. What is the meter resistance of a 0–250 V voltmeter having an ohms-per-volt rating of 22 kΩ/V? (3)

4. How many cycles of a 1 MHz signal will show on the screen of a c.r.o. when the timebase scan period is 2 μs? (3)

5. Calculate the frequency of the signal applied to a c.r.o. when 10 cycles is shown in a scan period of 1 ms. (3)

6. Calculate the ohms-per-volt rating of a 0–200 V meter using a 1 mA f.s.d., 20 Ω p.m.m.c. movement. (5)

7. Draw two diagrams showing how a wattmeter may be connected to a load. No explanation is required. (5)

8. In a c.r.o., state the function of the:
   (a) X- and Y-plates; (b) intensity control; (c) post-deflection accelerating anode. (5)

9. (a) State the characteristics of a good-quality voltmeter.
   (b) A voltmeter is used to measure the p.d. across a 10 kΩ resistor which is connected in series with a 20 kΩ resistor, the combination being supplied with 150 V d.c. The voltmeter rating is 10 kΩ/V and the range is 0–100 V. Calculate the p.d. across the 10 kΩ resistor before and after the voltmeter is connected in circuit. (14)

10. (a) Describe the two methods of connection of a dynamometer wattmeter. Compare the accuracy of the methods of connection.
    (b) A dynamometer wattmeter has a voltage-coil resistance of 8 kΩ and a current-coil resistance of 0.5 Ω. It is connected with its voltage coil in parallel with a 250 Ω load supplied from 200 V d.c. Calculate the wattmeter reading and the true power in the load. (14)

11. Describe the essential features of a cathode ray tube and explain how it may be used to indicate and measure voltage waveforms. (14)

12. Describe the d.c. potentiometer method of measuring an unknown e.m.f. In the description, explain how such a potentiometer can be calibrated to be virtually direct-reading. (14)

13. (a) Draw a diagram showing the basic arrangement of a Wheatstone bridge and explain how it may be used to measure resistance.
    (b) In such a bridge the ratio arms were of resistance 20 kΩ and 10 kΩ, one end of the 20 kΩ arm being connected to the same point as one end of the variable-resistor arm. Balance is attained when the variable resistor has a value 2200 Ω. What is the value of the resistance of the remaining arm? Draw a diagram of the circuit. (14)

Answers

EXERCISE 8

1. 9.975 kΩ
2. 0.5051 Ω
3. 1970 Ω/V
4. 1000 Ω/V
5. 0.02 A reduction
6. 19.53 V after; 20 V before
7. 0–1 V: 957 Ω. 0–1 A: 0.025 Ω
   0–10 V: 9975 Ω. 0–5 A: 0.005 Ω
   0–100 V: 99975 Ω. 0–20 A: 0.00125 Ω
8. Voltage coil across supply: wattmeter reading 414.46 W, true power 412.26 W
   Voltage coil across load: wattmeter reading 418.37 W, true power 412.19 W
9. (a) 198.22 W (b) 201.79 W
10. 897.47 W

Measuring instruments and measurements 175

11. (a) 20 (b) 200 (c) 20 000
12. A
13. 6.36 cm
14. 1.273 V
15. 195 cm
16. 57.2 Ω
17. 1/10
18. (a) 1/1000 (b) 1/1 (c) 1000/1

SELF-ASSESSMENT EXERCISE 8

Marks

1. Voltage across movement at f.s.d. = $1 \times 10^{-3} \times 15 = 0.015$ V (1)
Voltage across multiplier at f.s.d. = $100 - 0.015 = 99.985$ (1)
Multiplier resistance (f.s.d. voltage divided by f.s.d. current)
$$= \frac{99.985}{10^{-3}} = 99.985 \text{ k}\Omega \quad (1)$$

2. P.D. across movement at f.s.d. = $10 \times 10^{-3} \times 25 = 250 \times 10^{-3}$ V (1)
Current in shunt = $1 - (10 \times 10^{-3}) = 0.99$ A. (1)
$$\text{Shunt resistance} = \frac{250 \times 10^{-3}}{0.99} = 0.2525 \, \Omega \quad (1)$$

3. Meter resistance = 250×22 kΩ = 5.5 MΩ (3)
4. Period of 1 MHz signal = 1 μs; *two* cycles will be shown in 2 μs (3)
5. 10 cycles in 1 ms; 1 cycle in 0.1 ms, i.e. 0.1×10^{-3}s
$$\text{Frequency} = \frac{1}{0.1 \times 10^{-3}} = 10 \text{ kHz} \quad (3)$$

6. Total meter resistance (f.s.d. voltage/current) = $\frac{200}{10^{-3}} = 200$ kΩ (3)
$$\text{ohms/volt} = \frac{200 \text{ k}\Omega}{200 \text{ V}} = 1 \text{ k}\Omega/\text{V}. \quad (2)$$

7. Diagrams as fig. 8.13 ($2\frac{1}{2}$ each)
8. (a) X-plates provide horizontal deflection; Y-plates provide vertical deflection. (1)(1)
   (b) The intensity control controls the brightness of the trace on the screen. (1)
   (c) The post-deflection anode accelerates the beam, allowing increased brightness but not at the expense of deflection sensitivity. (2)
9. (a) It should not affect the circuit into which it is connected and should therefore have as high a resistance as possible. (2)
   (b) The total resistance is $(10 + 20)$ kΩ, i.e. 30 kΩ. (2)
   Voltage across the 10 kΩ resistor (without the voltmeter)
$$= \frac{10}{30} \times 150 \text{ V} = 50 \text{ V} \quad (2)(1)$$

Voltmeter resistance = 100×10 kΩ i.e. 1 MΩ. (1)

This is in parallel with the 10 kΩ resistor making the resistance of the combination
$$\frac{10^6 \times 10^4}{10^6 + 10^4}, \text{ i.e. } 9901 \text{ k}\Omega \quad (2)$$

Total resistance = $(20 + 9.901)$ kΩ = 29.901 kΩ. (2)

P.D. across 10 kΩ resistor with voltmeter in circuit is

$$\frac{9.901}{29.901} \times 150, \text{ i.e. } 49.67 \text{ V} \qquad (2)(1)$$

10. (a) Description should include diagrams (fig. 8.13), and explain that there is a loss with each method of connection. (2 each)
    (b) Resistance of load in parallel with voltage coil

$$= \frac{250 \times 8000}{8250} = 242.42 \text{ }\Omega \qquad (2)$$

Circuit resistance $= (242.42 + \text{current-coil resistance}) = 242.92 \text{ }\Omega$ (1)

Supply current $= \dfrac{250}{242.92} = 1.029$ A. (1)

P.D. across current coil $= 1.029 \times 0.5 = 0.514$ V (1)

P.D. across voltage coil/load $= 250 - 0.514 = 249.48$ V (1)

Wattmeter reading $= 249.48 \times 1.029 = 256.72$ W (2)

True power in load $= \text{voltage}^2/\text{resistance} = 249.48^2/250 = 248.96$ W (1)(1)

11. Essential features must include:
    (i) Electron gun, heater, cathode and modulator (3)
    (ii) Focus; (iii) deflection; (iv) post-deflection anode (1 each)

    Description of use must include description of:
    (i) Screen trace, no voltage at X- or Y-plates (1)
    (ii) Screen trace, signal at X-plates alone and at Y-plates alone (2)
    (iii) Timebase waveform (1)
    (iv) Screen trace, signal at Y-plates, timebase at X-plates (1)
    (v) How signal magnitude is obtained (deflection sensitivity, etc.) (3)

12. Description must include:
    (i) Circuit diagram showing supply, slide-wire, test e.m.f. (2)
    (ii) Why a balance can be obtained (opposing effects of test cell and supply) (2)
    (iii) Insertion of standard; balance (2)
    (iv) Insertion of unknown; balance (2)
    (v) Calculation of unknown using potentiometer equation. (3)

Direct reading: balance length with standard in circuit is adjusted to be an integral multiple of standard cell e.m.f. An example should be given. (3)

13. (a) Diagram, see fig. 8.18. (3)

Explanation of why a balance can be obtained (p.d. across arms joined to same point of supply being the same and thus potential at each side of galvanometer being the same). (2)

Derivation of balance equation. Explanation of use of ratio arms, variable resistor and unknown, *and* how a suitable choice of ratio can give better accuracy. (3)(2)

(b) Diagram: fig. 8.18 with $R_1 = 10$ kΩ, $R_2 = 20$ kΩ, $R_3 = 2200$ Ω. (3)

$$R_4 = R_3 \times \frac{R_1}{R_2} = 2200 \times \frac{10}{20} = 1100 \text{ }\Omega \qquad (2)$$

Index

absolute permeability, 41, 53
absolute permittivity, 23, 32
alloy junction transistor, 138
alternating current, 78 *et seq*.
   average value, 84, 96
   defined, 78
   form factor, 89
   instantaneous value, 83
   peak/peak value, 83
   phase, 90
   r.m.s. value, 86
   sinusoidal, 80
alternating voltage, 78 *et seq*.
   average value, 84, 96
   defined, 78
   form factor, 89, 97
   instantaneous value, 83, 96
   peak/peak value, 83, 96
   phase, 90
   r.m.s. value, 86, 96
   sinusoidal, 80
ammeters, 152, 171
ampere, 1
amplitude, 81
anode, 134, 136
apparent power, 120, 124
arsenic, 132
avalanche effect, 134, 136
average values, 84

back e.m.f., 64
bar magnets, 60
barrier potential, 134
base, 138
bias,
   diode, 134
   transistor, 138
bipolar transistor, 138 *et seq*.
breakdown voltage, 19
bridge measurement, 167, 172

CR series circuit (a.c.), 110
capacitance, 20, 31
capacitance reactance, 106 *et seq*., 122

capacitor, 20, 31
   connections, 26
   energy, 28
   types, 29
cathode, 134, 136
cathode ray oscilloscope, 161
   et seq., 172
   tube, 161
charge, 16, 31
chokes, 66
coercive force, 44, 54
collector, 138
common base circuit, 142, 143
common collector circuit, 142, 143
common emitter circuit, 142
contra-wound springs, 149
control, instrument, 149
current, 1, 10
cycle, 81, 96

damping, instrument, 150
d'Arsonval instrument, 149
d.c. potentiometer, 168
deflection, c.r.o., 161
depletion zone, 134
diamagnetic materials, 42, 54
dielectric, 20, 32
dielectric constant, 24, 32
dielectric strength, 25, 31
diodes, 130 *et seq*.
doping, 132 *et seq*.
dynamometer wattmeter, 155
   connections, 158
   error, 158

efficiency, transformer, 71
electric charge, 16, 31
electric field strength, 17, 31
electric flux, 17, 31
   flux density, 17, 31
electrodynamic instrument, 150
electromagnetic damping, 151
electromagnetic induction, 63
   et seq.

electromechanical instruments, 148 *et seq.*
electromotive force, 1, 10
electron gun, 161
e.m.f., 1, 10
emitter, 138
emitter follower, 143
energy
   capacitor, 28, 32
   inductor, 68, 73
error, instrument, 156 *et seq.*
extrinsic semiconductor, 132

Faraday, Michael, 63
ferromagnetic materials, 42, 54
field effect transistor (FET), 138, 144
field pattern, magnetic, 38
field strength, electric, 17, 31
   magnetic, 41, 53
flux, electric, 17, 31
   magnetic, 38, 53
flux density, electric, 17, 31
   magnetic, 40, 53
flyback, 164
focussing, c.r.o., 161
form factor, 89, 97
forward bias, 134, 139
frequency, 82, 96
   resonant, 114
full scale deflection, 149

gallium, 132
galvanometer, 167
germanium, 132

h_{fe}, 140
h_{FE}, 140
h parameters, 140
hertz, 82
hole, 132 *et seq.*
hole conduction, 139
hysteresis, 44, 54
   loop, 45, 54

impedance, 104, 110, 123
indium, 132
inductance,
   mutual, 69, 73
   self, 66, 72
induction 63 *et seq.*
inductive reactance, 105 *et seq.*, 122
inductor, 66
input impedance, 143
instantaneous value (a.c.), 83, 96
instrument error, 156
instruments, *see under* type
intrinsic semiconductor, 132
isolation transformer, 70

junction p.d., 134

Kirchhoff's Laws, 3, 4, 11

leakage current, 140
Lenz, 64
LR series circuit (a.c.), 109

magnet, bar, 60
magnetic field strength, 41, 53
magnetic flux, 40, 53
   flux density, 40, 53
   hysteresis, 43, 54
   saturation, 43
magnetising current, 70
magnetomotive force, 39, 53
m.m.f., 39, 53
moving iron instrument, 148
multiplier, 152
mutual inductance, 69, 73

n-type semiconductor, 132
null methods, measurement, 167, 172

ohm, 2, 10
Ohm's Law, 2, 10
ohm-meter, 153
ohms-per-volt, 156
output impedance, 143

p-type semiconductor, 133
parallax, 161
parallel connection,
   capacitors, 26, 32
   resistors, 2, 10
paramagnetic materials, 42, 54
peak/peak value (a.c.), 83, 96
peak inverse voltage, 135
pentavalent materials, 132
period, 81, 96
periodic time, 81, 96
permanent magnet, moving coil instrument, 149, 171
permeability,
   absolute, 41, 53
   free space, 41, 54
   relative, 41, 54
permittivity,
   absolute, 23, 32
   free space, 24, 32
   relative, 24, 32
persistence, c.r.o., 163
phase (a.c.), 90
phase angle, 111
phasors, 91 *et seq.*, 97
   addition, 94
   components, 95
   resolution, 95

Index

planar transistor, 138
potential, 18
potential difference, 1, 18, 10, 31
potentiometer, d.c., 168
power, 1, 10
   a.c. circuits, 118 *et seq.*, 124
power factor, 119, 124

Q-factor, 116, 124

reactance, 104 *et seq.*, 122
   capacitive, 106, 122
   inductive, 105, 122
reactive volt-amperes, 119, 124
rectification, 136 *et seq.*
rectifier instruments, 154
relative permeability, 41
relative permittivity, 24
reluctance, 46, 54
remanent value, 44, 54
resistance, 2, 10
resistors, 2 *et seq.*
resolution of phasors, 95
resonance, 114, 123
resonant frequency, 114, 123
reverse bias, 134, 139
r.m.s. value (a.c.), 86, 96

saturation, magnetic, 43
scan p.d., 164
self inductance, 66, 72
semiconductor, 130
   diode, 134 *et seq.*
series a.c. circuits, 109 *et seq.*
   CR, 110
   LCR, 114
   LR, 109
series connection,
   capacitors, 26, 32
   resistors, 2, 10
series resonance, 114
set zero control, 154
single phase circuits, 104 *et seq.*
silicon, 130
sinusoid, 80
slide wire, 168

square-law scale, 149, 171
standard cell, 168
suppressor circuit, 66
sweep voltage, 164
sync control, 166
synchronisation, 165

taut band control, 149
tesla, 40
thermal runaway, 141
timebase, 164
transformer, 69 *et seq.*, 73
transients, 20
transistor, 137 *et seq.*
trigger control, 166
trivalent materials, 132

valency, 131
viscous damping, 151
volt, 1
volt-amperes, 120
   reactive, 119
voltage gradient, 18
voltage magnification, 116, 124
voltmeters, 152, 171

wattmeter, 155
   connections, 158
   errors, 158
waveform
   definition, 80 *et seq.*
   display, 163
   frequency, 82
   graphical addition, 90
   period, 81
   values, 83 *et seq.*
Wheatstone Bridge, 167, 169, 172

X-plates, 163
X shift, 163

Y-plates, 163
Y shift, 163

Zener effect, 134, 136